Relativity of Artificial Intelligence

Hasan Sadik

Published by Hasan Sadik, 2023.

RELATIVITY OF ARTIFICIAL INTELLIGENCE

First edition. October 24, 2023.

Copyright © 2023 Hasan Sadik.

ISBN: 979-8215725603

Written by Hasan Sadik.

Relativity OF
Artificial intelligence

The Third Theory of relativity

BY

HASAN SADIK

RELATIVITY
OF
ARTIFICIAL INTELLIGECE

proofreading and editing by

Dr. Kim

Hasan Sadik

Berlin – Germany

physicsshock@hotmail.com

TEL: +4917641724207

INTRODUCTION

Today, artificial intelligence is feeds us with accurate and renewable information every day, and it supports us with creative ideas, from essays to guides, to webinars, to deep learning. The emergence of new horizons and sophisticated technology has increased our knowledge and rapidity. Today, we do not have to do much more to get to the truth. We have become able to recognize the tools everyone uses, take inspiration from some amazing projects, and look for experts who can help. What I found in relativity, interesting and surprising, leads to a new relativism in physics, which I called AI relativism.

"Einstein" was not the first to apply the laws of relativity; rather, it was the Italian astronomer "Galileo Galilei", who was the first to use the laws of relativity in the mid-sixteenth century in his astronomical measurements between stars, planets, and their moons. Sir "Newton" set the laws of relativity and its postulates, mentioning them in his book; "Mathematical Principles of Natural Philosophy". (Philosophiæ Naturalis Principia Mathematica).

After "Maxwell" published the electromagnetic theory, applying the laws of relativity to Maxwell's equations became difficult, including the electromagnetic wave's speed. This compelled scientists to search for a mathematical equation to measure the speed of the Earth around the sun. The controversial Michelson-Morley experiment was the only option that did not yield the desired result. The experiment also marked a significant turning point in the history of physics. At that time, there was a need to develop new transformations consistent with the motion of light when using a moving observer in equations.

The Dutch mathematician Hendrik Antoon Lorentz was able to derive the correction factor for the distance travelled in the direction of motion. He also managed to derive new transformations that could be applied to Maxwell's equations. These transformations resulted in the contraction of the length of a moving body in the direction of motion as its speed increased. As a result, this led to the emergence of a new physics called modern physics, which was founded based on Einstein's theory of relativity.

Einstein noticed this equation and considered that the Lorentz factor could be applied to mass, time, velocity, and energy. Moreover, Lorentz derived the equation for length contraction in collaboration with the Irish physicist Fitzgerald.

Einstein published his theory in 1905 with the five equations that indicate changes in physical properties with changes in velocity. He called this theory the Special Theory of Relativity. The equations of relativity did not align with reality and could not be proven in any scientific experiment since it is impossible to reach speeds close to the speed of light. This fact reassured Einstein, as it meant that his theory of relativity, which had sparked widespread controversy in the scientific community, could not be disproven.

his theory gained fame from its strange, astonishing, and contrary-to-reality results, which raised doubts about its validity.

Many physicists and intellectuals tried to refute Einstein's theory. Still, their efforts were in vain because their publications did not provide a sufficient alternative, and it was impossible to prove the failure of the theory of relativity experimentally. Although the shortcomings of Einstein's relativity were apparent to many, the proof of its failure was not found in the arguments of the sceptics.

Here, I would like to mention what the British scientist Stephen Hawking wrote in his book *"Brief Answers to the Big Questions"*, where he said: "If someone were to apply for a grant to study time travel, they would be immediately dismissed, according to the American website; Science Alert. This is because the question of the possibility of time travel is an extremely serious one and cannot be addressed scientifically because any research that proves the impossibility of time travel is, in reality, a refutation of the theory of relativity, one of the most important principles of modern physics."

While proving the failure of relativity may not require much effort or research, it is important to note that the conditions under which relativity holds suggest that its principles only apply to objects moving according to Newton's first law. Hitherto, no one has been able to modify or correct Einstein's relativity. Many physicists hesitate to propose alterations or amendments to established physical laws, even before examining them, due to the wide-ranging fame associated with Einstein and the perception that science has reached a pinnacle with his theory.

However, in today's world, with the help of computers, the internet, and artificial intelligence, we have the tools to explore scientific possibilities.

Here, I recall the scientist Max Planck, who desired, after he graduated from high school in Munich in 1874, to attend the university in the physics department. However, his teacher, Philipp Von Jolly, advised him to choose a field other than physics since renowned scientists had thoroughly developed it, and Max could add nothing or modify it except to fill in some gaps in the existing knowledge. Nevertheless, Max insisted on studying physics and made significant contributions beyond what his teacher expected.

He established a new theory, Quantum Theory, which forms the basis of quantum physics. We hope our scientists, professors, and students will adapt. Science does not stop at any stage and will continue to develop, evolve, and undergo modifications, and might never reach perfection.

I have put all my effort into this book, explaining some of the insignificant details in a simplified manner so that the ideas can reach even the educated reader and physics enthusiasts to facilitate the comprehension of challenging topics and complex equations for all readers.

Chapter I

Newtonian Classical Relativity

8

Galilean Transformations

Indeed, both intellectuals and the general public must have heard a lot about Einstein's theory of relativity, which has fundamentally altered many concepts in physics. This theory has astonished everyone, including the scientists themselves, with its ideas about time travel, the contraction of moving bodies, and various other imaginative thought experiments. Its results often defy logic and appear to contradict the laws of nature.

However, Einstein was not the first to establish the principles of relativity. Those principles emerged in the time of the Italian astronomer Galileo Galilei, who used relative transformations to describe the motion of celestial bodies. However, he did not publish them or formulate a final mathematical expression.

However, Sir Isaac Newton, in the 17th century, was the one who laid the first principles of relativity theory. These principles indicate that space and time are constant and absolute quantities that do not change with varying velocity or mass. He formulated two equations for this purpose: the first expresses the coordinates of the stationary observer, and the second expresses the coordinates of the moving observer.

Connecting the coordinates of one observer to those of the other is known as relativity, as motion is relative between observers. These equations were named Galilean transformations, and they are two simple equations that can be mathematically expressed as follows:

$$x = vt` + x` \quad \dots\dots\dots\dots (1)$$
$$x` = x - vt \quad \dots\dots\dots\dots (2)$$

Equation (1) represents the distance between the stationary observer and the event, equal to the distance the observer covers towards the event, plus the remaining distance between the moving observer and the event.

Equation (2) represents the distance the moving observer perceives to reach the event. This distance equals the distance between the stationary observer and the event minus the distance the moving observer covers.

We can express the coordinates of these two equations as follows:

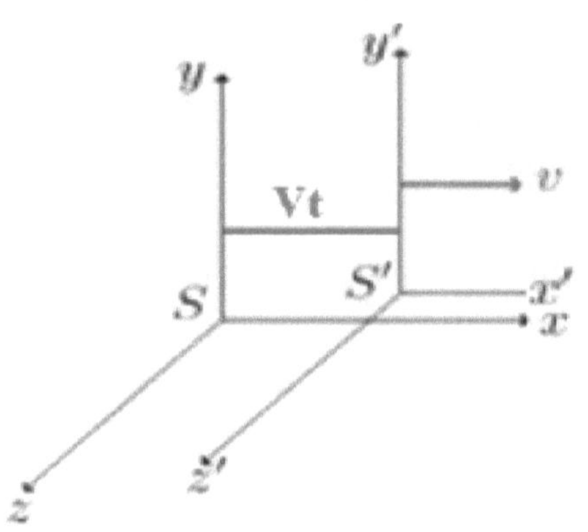

S: Represents the stationary frame.

   S`: Represents the moving frame.

   V: Represents the velocity of the moving frame.

   t : Represents time.

One of the fixed postulates in the relative relationship between two frames is that Galilean transformations apply only to objects subject to the first law of motion, the Law of Inertia. This law states that a stationary object, incapable of self-motion, will remain at static unless acted upon by an external force that moves it from its position. Similarly, a moving object, incapable of self-motion, will continue to move in a straight line at a constant velocity unless influenced by an external force that alters its speed or direction. Under these basic conditions, we can dispense with the y and z axes coordinates because the laws of relativity do not apply to motion occurring in more than one axis.

One of the constants of the relativity principle is that the physical laws used by an observer in the stationary frame remain unchanged when used by an observer in the moving frame. To illustrate this concept with a practical experiment, let us assume a person is holding a ball while standing on a train station platform, waiting for a train to arrive.

Suppose they want to determine the acceleration due to gravity. All they need to do is to toss the ball upwards to a height of one metre. They will find that the ball returns and falls into their hand with an acceleration of approximately 9.8 metres per second, even though the Earth is not stationary and is moving at a speed of around one thousand kilometres per second relative to the Absolute stillness in the universe.

After the person boarded the train and travelled a distance until it reached a constant speed, they wanted to repeat the same experiment by tossing a ball the same height. They found that the ball returned to their hand at the same speed and acceleration as when they were standing on the platform. This means that the outcome of the physical process remains constant, as if the train were stationary and not moving. This is the principle of the constancy of the physical laws in all different inertial frames.

In other words, the outcome of the physical process does not change with changes in speed, assuming that the Earth is stationary, and the train is the one in motion. Based on the principles of relativity, we can consider the train as stationary and the Earth as moving.

nevertheless, the physical law remains constant and does not change with changes in the frame, regardless of its speed or cosmic velocity.

To conclude this in a mathematical equation, we can apply the transformations to the law of force, as we illustrate below:

F = ma

Force equals mass multiplied by acceleration.

and by analysing the following equation:

F = ma

F = m.dv/dt

$$F = m.\frac{d^2x}{dt^2}$$

When applying Galilean transformations to the law of force when used by an observer in motion, the law takes the following mathematical form:

F` = ma`

By analysis of the equation used by the observer in motion when observing force, the equation takes the following mathematical form:

$F` = m.d^2x`/dt^2$

x` = x − ut

Where:

F` - represents the force observed by the observer in motion.

m -represents mass.

a` - represents the acceleration observed by the moving observer.

X - represents the distance as observed by the stationary observer.

X` - represents the distance as observed by the observer in motion.

U - represents velocity.

T - represents time.

And by substituting the value of x`:

F` = m.d/dt[dx/dt - d/dt.(ut)]

F` = m.dt/d(dx/dt - u)

$$F` = m.\frac{d^2x}{dt^2}$$

Here, we find that the result of the acceleration law used by the observer in motion, is the same as that result of the stationary observer. This confirms the consistency of the physical law used by all observers, regardless of their velocities.

Therefore, the transformations become unnecessary or meaningless, as each observer can use the same law without needing the transformations.

The linking between coordinates of two observers in different frames, are related to the distance and the relative velocity between them nothing more.

The Electromagnetic Theory

Classical Newtonian Relativity prevailed for about two hundred years until the British scientist Maxwell published the equations of electromagnetic theory between 1861 and 1862.

These four fundamental differential equations describe the behaviour of the magnetic and electrical waves and their influence on each other. Maxwell, in an astonishingly brilliant manner, determined the speed of the wave resulting from the union of the magnetic wave with the electrical wave. He did this using the magnetic permeability and the electric permittivity, which he called the electromagnetic wave.

The following illustration shows the shape of the electromagnetic wave:

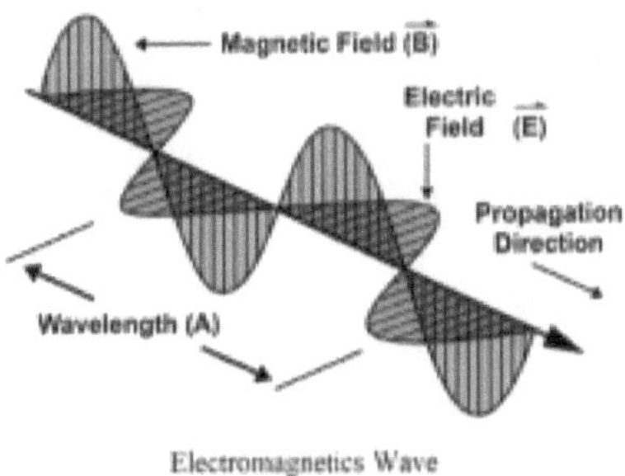

Electromagnetics Wave

The electromagnetic theory consists of four fundamental equations, which are:

1. - Electrical Flux Equation (Gauss's Law):

$$\nabla \cdot E = \rho/\varepsilon_0$$

Where:

∇: is the divergence of the electrical field (E)

E: is the Electrical Field

P: is the charge density

ε_0: is the vacuum electric permittivity constant

1. - The Magnetic Divergence Equation (Gauss's law for magnetism):

$$\nabla \cdot \mathbf{B} = 0$$

Where:

B: is the magnetic field vector

1.   - Faraday's law of electromagnetic Induction:

$$\nabla \times \mathbf{E} = - dB/dt$$

Where:

t: is time.

dB/dt is the time derivative of the magnetic field.

4 - The magnetic curl equation (Ampère- Maxwell's law).

$$\partial^2 E/\partial \diamond^2 = (\mu_0 \epsilon^2 \partial .(_0 E/\partial t^2$$

Where:

μ_0: Magnetic permeability constant of vacuum

$\epsilon 0$: Electric permittivity constant of vacuum

dE/dt: Time derivative of the electric field.

The speed of the electromagnetic wave in a vacuum is a constant value and equal to the reciprocal of the square root of the magnetic permeability multiplied by the electric permittivity.

$$\text{Constant} = \frac{1}{\sqrt{\mu 0 \, \epsilon 0}}$$

So, the fourth equation takes the following mathematical form:

$$\partial^2 E/\partial \lozenge^2 = 1/c^2 \, . \, \partial^2 E/\partial t^2$$

Where:

> C: is the constant of the velocity of the electromagnetic wave in a vacuum.

The speed of the electromagnetic wave is considered the fastest thing in the universe, and nothing cannot match it or even reach it. It travels in a vacuum at a speed of 299,792,458 metres per second.

Maxwell was surprised when he discovered that the speed of electromagnetic waves in a vacuum is exactly equal to the speed of light, which many scientists have experimentally measured.

Therefore, Maxwell considered that light must be an electromagnetic wave.

Scientists wondered how electromagnetic waves could propagate in a vacuum, given that sound waves propagate in matter, not in a vacuum, as waves involve vibrations in a material medium. Hence, they assumed the vacuum must contain a substance that carries the electromagnetic wave.

Maxwell proposed the existence of a substance that fills the vacuum and called it "Ether." This ether carries light waves throughout the universe and is considered the state of absolute rest in the cosmos.

The electromagnetic wave can propagate through anything, whether it is a gaseous, liquid, solid material, or a vacuum. The speed of electromagnetic waves in any physical medium can be calculated using the law that states that the speed of electromagnetic waves is equal to the speed of the wave in a vacuum divided by the square root of the electric permittivity constant of the transmitting medium.

The speed of light has become an expression of the speed of the electromagnetic wave. Since the speed of light has a constant value relative to the ether, when applying Galilean transformations to

Maxwell's fourth equation in the moving frame, which in this case is the Earth moving in the ether around the Sun, the equation no longer maintains the same mathematical form of the law. Below is the mathematical equation that represents this:

$$\partial B/\partial \Diamond = 1/c^2 \cdot \partial E/\partial t \dots\dots\dots(1)$$
$$\partial B`/\partial \Diamond` = 1/c^2 \cdot \partial E`/\partial t` \dots\dots(2)$$

After analysing the equation used by the moving observer through lengthy and complex procedures, it is found that it does not match the first equation used by the stationary observer. The equation used by the moving observer takes the following mathematical form:

$$\frac{\partial B'}{\partial X'} = \frac{1}{c^2} * \frac{\partial E'}{\partial t'} - \frac{1}{c^2} * \left[u * \frac{\partial E'}{\partial t'} {}_{-u} * \frac{\partial B'}{\partial t'} + u * \frac{\partial B'}{\partial X'} \right]$$

In order to apply the Galilean transformations to Maxwell's equation, we need to know the speed of the Earth in the ether, which has become a part of the mathematical equation. We know that the speed of the Earth in its orbit around the Sun, according to well-established astronomical measurements, is thirty kilometres per second. Therefore, scientists attempted through various methods to determine the speed of the ether that confronts the Earth during its orbital motion around the Sun, as this represents the Earth's speed we need in the equation when the law is used by an observer in motion along with the Earth in the ether. The diagram below illustrates the Earth's motion around the Sun in the context of the ether.

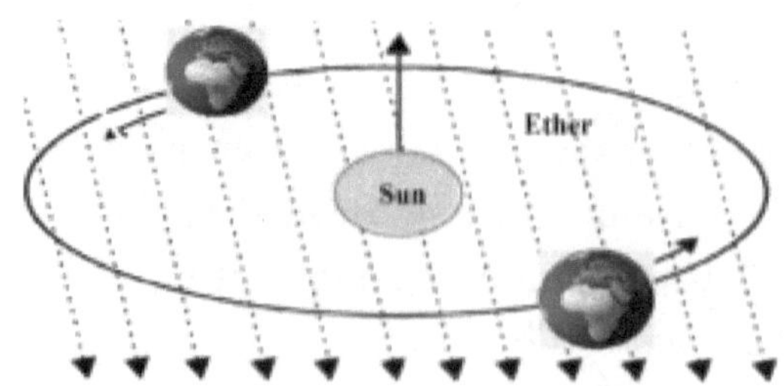

The best attempts and most accurate experiments were conducted by the American scientists Michelson and Morley in 1886. These experiments are among the most significant in physics. Despite their significant importance, scientists struggled to explain their strange and unexpected results. Therefore, we will explain the experiment in detail with precise and new scientific details, as outlined below.

The Michelson and Morely Experiment

In 1886, scientist Albert A. Michelson designed a device, which was a square wooden piece mounted on a rotating base. In the centre of this device, he placed a half-silvered mirror that separated the incoming light beam from only the source and directed the light in two perpendicular directions. One of the beams reflected off the mirror and headed towards a second reflecting mirror, while the other beam passed through another reflecting mirror. It was designed so that one of the beams was parallel to the axis of the Earth's rotation around the Sun, and the other beam was perpendicular to it.

In this setup, one of the beams must be affected by the Earth's motion through the ether, causing its speed to be greater than the other beam. Then, the two beams are recombined and directed onto a detector. If there is any difference in the speed of either of the beams, it will affect the interference pattern between the two beams on the detector, creating bright and dark regions.

The bright fringes move in a particular direction. Using this setup, Michelson was able to measure the difference in the speed of light in the two directions. This allowed him to measure the Earth's velocity through the ether, and consequently, apply the Galilean transformations to Maxwell's equations.

The following form represents the description of the experience:

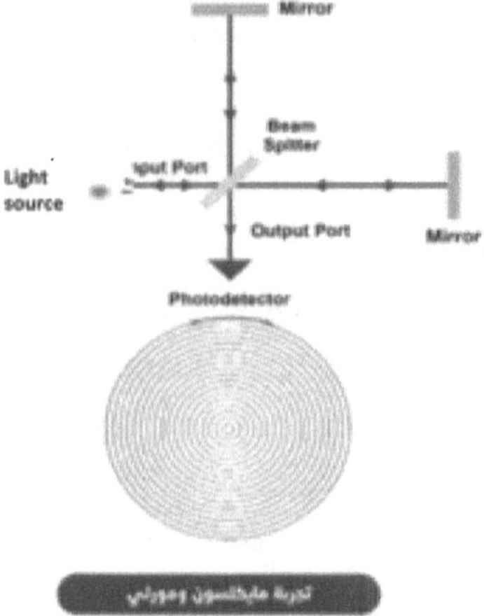

If we assume the absence of the ether, then the beams in both directions would travel the same distance. Therefore, both beams would require the same amount of time to reach the detector. In this case, the time taken in both directions (going and returning) would be:

$$T = \frac{D}{C} + \frac{D}{C} = \frac{2D}{C}$$

In the presence of the ether, the measurements of the beam facing the ether differ from the measurements of the beam moving perpendicular to the ether. The equations become as follows:

The beam moving perpendicular to the ether, in its outbound journey to the mirror, will be perpendicular to the ether. On its return to the detector, the beam will still be perpendicular to the ether. The total time can be described by the following equations:

$$T_1 = \frac{D}{C\sqrt{1-u^2/c^2}} + \frac{D}{C\sqrt{1-u^2/c^2}}$$

$$T_1 = \frac{2D}{C\sqrt{1-u^2/c^2}}$$

As for the beam moving parallel to the ether, it will face the ether during its outbound journey to the mirror and will move along with the ether on its return to the detector or vice versa. The equations are as follows:

$$T_2 = \frac{D}{C-U} + \frac{D}{C+U}$$

$$T_2 = \frac{D(C+U)+D(C-U)}{(C+U)(C-U)}$$

$$T_2 = \frac{2DC}{C^2\left(1-\frac{U^2}{C^2}\right)}$$

$$T_2 = \frac{2D}{C\left(1-\frac{U^2}{C^2}\right)}$$

From the equations, it is evident that the arrival times of the two beams differ. The principle of the experiment relied on the idea of a practical experiment involving two boats moving in a flowing river, representing the ether. Both boats have the same speed and travel the same distance. However, one of them crosses the river to the opposite bank and returns to the starting point, while the second boat moves upstream by the river's width and then returns to its starting point. Although the boats have the same speed and cover the same distance, the arrival time of the first boat is less than that of the second boat due to the river's motion. Scientists compared Michelson-Morley's experiment to the boat experiment, and the mathematical details of the boat experiment should align with Michelson-Morley's experiment, as shown in the following mathematical formulation:

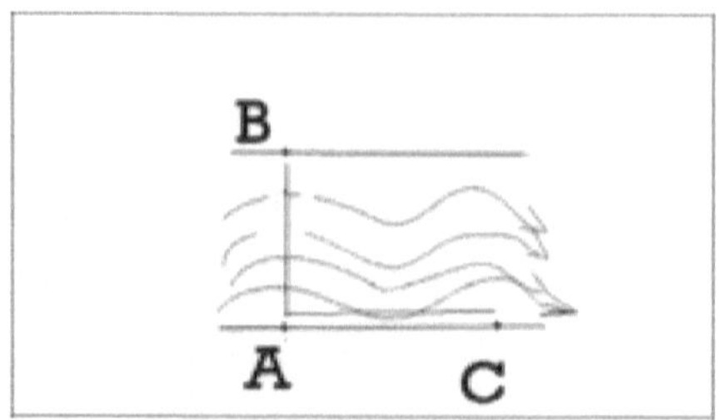

The total time taken by the first boat, which moves perpendicular to the current, is given by:

$$T_1 = \frac{2D/V}{\sqrt{1 - V^2/u^2}}$$

The time taken by the second boat, which moves with the current, going and returning, is given by:

$$T_2 = \frac{2D/V}{1 - \left(\frac{v^2}{u^2}\right)}$$

Where:

A: is the starting point of the two boats.

AB : is the width of the river.

AC : is the distance travelled by the boat parallel to the current.

T_1 : is the time taken by the boat crossing the width of the river.

T_2 : is the time taken by the boat moving with the current.

V : is the boat speed.

U : is the speed of the river current.

D : is the distance covered.

According to the results of the mathematical equations, the first boat will return to the starting point before the second boat because it takes less time than the second boat.

Michelson expected, based on this experiment, that there would be a displacement in light waves due to the difference in the speeds of the two beams in the experiment apparatus, typical of the two boats' motion. The light would appear on the detector in the form of interference fringes moving with a phase difference of one-third of a complete wave or with a time difference between the movements of the two beams, as illustrated below.

$$\Delta t = \sec (3 \times 10^{-17})$$

This is what Michelson accurately predicted before starting the experiment. Despite the precision sensitivity of this device, no difference between the speeds of the two beams was recorded, and the expected interference fringes displacement did not appear.

Some scientists pointed out the possibility that the timing of the experiment coincided with the motion of the Earth in the direction of the ether. Considering that the Sun also moves in the ether, some of them suggested conducting the experiment six months later.

Michelson repeated the experiment dozens of times, making improvements to the device over a period of ten years. He conducted the experiments at separate times of the day and in different seasons, but the result was consistent: there were no moving interference fringes. This indicates one of two possibilities. Either the Earth is stationary and not moving or rotating around the Sun, which is an unreasonable assumption, or the assumption of the existence of the ether is illogical. And this possibility was rejected by scientists, who insisted on the necessity of the existence of the ether to transmit light waves.

Michelson considered that the experiment did not achieve its goal of measuring the Earth's speed. It became evident that it did not match the boat experiment, and there was no longer any possibility of measuring the absolute speed of any object in the universe. All we can measure is the relative speed between two frames, both of which must be in motion, and there is no stationary frame in the universe.

However, the experiment confirmed that the speed of light is constant and equal in all directions on the surface of the Earth. It is as if the Earth is a stationary frame with respect to light, just like it is a stationary frame for sound and all the materials present on the Earth's surface, its surroundings, and everything within its environment.

The idea behind the experiment was not bad at all, and its results were tremendously valuable. However, the design of the experiment had some shortcomings. The scientist Michelson did not take into account the motion of the Earth within the movement of the Sun around the centre of the galaxy, which is about 220 kilometres per second. Additionally, the Earth is moving at a speed of approximately eight hundred kilometres per second with the motion of the galaxy in an unknown direction. This means that the speed of the Earth in the ether is more than one thousand kilometres per second relative to the ether, not just thirty kilometres per second. Furthermore, the unknown direction of this motion would prevent the experiment from matching the boat experiment in any case.

Until 1929, scientists believed that the universe consisted only of our galaxy, the Milky Way, which contains about 200 to 300 billion stars. However, the discovery of the speed of light provided scientists with new tools. Measurements conducted by the astronomer Edwin Hubble between 1927 and 1929, using his modern telescope and calculations of the speed of light, revealed the existence of independent galaxies. Hubble found a galaxy known as Andromeda, which is approximately 2.5 million light-years away from us. This discovery contradicted the expectations of Michelson's experiment.

Finding an explanation for the experiment's results became essential, and scientists were puzzled. The only explanation came from the Irish scientist Fitzgerald, was a proposal that changed the course of physics incredibly.

Chapter II

Einstein's relativity

29

Fitzgerald's principle of contraction

Michelson initially disregarded the experiment, believing that it did not achieve its intended purpose. However, upon verifying the device's construction and repeating the experiment multiple times by various scientists, a conclusion was reached stating that the speed of light is not related to the speed of the source.

The experiment was not a failure, as Michaelson described, due to his inability to determine the speed of the Earth in the ether. Instead, it was one of the most remarkable experiments in physics, as it demonstrated the behaviour of light. To interpret the results of the Michelson-Morley experiment and explain its significance, there were three important options, as follows:

The first option is the proposal of "Fitzgerald", which was adopted by "Lorentz". The Irish physicist, George Fitzgerald, attempted to reinstate the idea of the existence of the ether and provide an explanation for the unexpected results of the experiment. He introduced the concept of "length contraction" in the direction of motion relative to the ether. The idea was based on the boat analogy. To make the Michelson-Morley experiment consistent with the boat analogy, it was assumed that the beam of light facing the ether covers a shorter distance than the beam perpendicular to the ether. This is due to the contraction of the Earth in the direction of its motion relative to the ether. As the speed of the Earth increased, the distance contracted even more, and the beam of light required less time to cover that distance. As a result, the times for the two beams to reach the detector became equal.

The assumption is that the contraction only occurs in the direction opposite to the ether and with respect to it, not in any other random direction. Otherwise, the idea of contraction would be inappropriate and illogical, and its application would contradict the principle proposed by Fitzgerald.

The Dutch scientist Lorentz relied on Fitzgerald's principle and attempted to explain these results within the framework of traditional physics. The most important part of the analysis was that the spherical shape flattens when it moves in the direction of its motion, and the faster it is, the more the matter flattens to some extent when it moves along its path. Fitzgerald used this explanation to interpret the results of Michelson and Morley's experiment, confirming the existence of the ether. He also explained its effect on the contraction of distance in the same direction as the motion of light, which contracts by a precise amount necessary to cancel the delay caused by the ether current facing the Earth's motion, appropriate to its speed.

The following figure illustrates the effect of speed on the spherical shape, as described by Fitzgerald, which is assumed to be described only in the direction facing the ether.

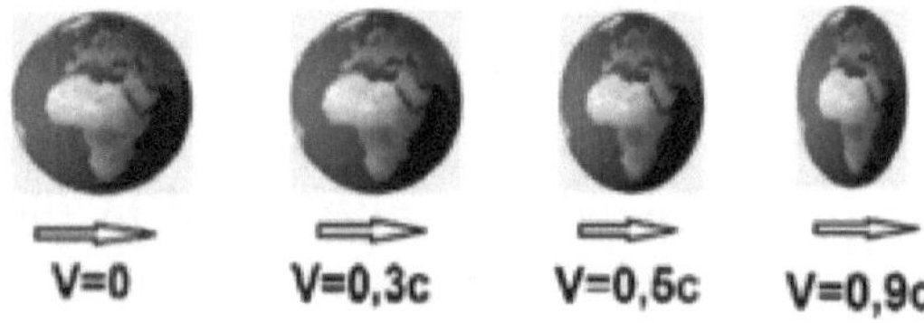

In the next paragraph, we will learn about Lorentz transformations according to the Fitzgerald principle, after their detailed analysis and an explanation of their mathematical form in a new scientific approach, relying on artificial intelligence that has provided us with sufficient evidence for the accurate interpretation of Lorentz transformations and the derivation of alternative transformations.

The second option is that the distance remains constant and unchanged due to the absence of the hypothetical ether substance, but the time of the light ray in the direction of motion changes. It slows down by the same amount as the difference between the times of the two rays due to the constant speed of light. In this case, the arrival time of the two rays at the detector becomes equal. This was the choice of Einstein, who adhered to it and denied the existence of the ether, which would lead to the contraction of distance in the direction of motion.

To explain and clarify Einstein's choice, let us assume that a pulse of light is emitted from a stationary light source in an absolutely motionless frame in a vacuum. In that case, that pulse will travel a distance of 299,792,458 m/s. However, a pulse of light emitted from a light source in a frame moving at half the speed of light will still cover the same distance, but it will take approximately 1.1547344 seconds to do so.

While Lorentz relied on the two-boat analogy, Einstein based his theory on an experiment known as "Einstein's Train". This experiment involves the motion of a light beam within a moving train car, which is analogous to the motion of a ball thrown upward and downward inside the train car, as observed by an observer standing at a train station and watching the motion. If length contraction occurs simultaneously with time dilation in one experiment, the result of the equation in the moving frame would not differ from the result of the equation in the stationary frame. We will explain this in detail in the next chapter.

As for the third option, it was overlooked by Lorentz, Einstein, and even Michelson, the creator of the experiment. None of the scientists addressed it. This option is that the experiment confirmed the validity of the principle of relativity, which asserts that the laws of physics remain constant in all uniformly moving reference frames and are not affected by changes in the moving frame speed, including the laws of electromagnetism. This is our choice, and it is the correct one. In this view, the Earth becomes a stationary frame for light, just as it is a stationary frame for sound or any other object on its surface or in its surroundings. This is what an observer on the surface of the Earth perceives. We will explain this option in detail in Chapter Three.

After the Michelson and Morely experiment failed to determine the speed of the Earth through ether, the search for new transformations became scientists' most important thought. The best transformations were those derived by Lorentz based on the principle proposed by Fitzgerald. Below is an explanation of these peculiar transformations after a thorough study.

The Absurd Lorentz Transformations

Lorentz relied on Galilean transformations and applied them to the Michelson-Morley experiment following Fitzgerald's principle in formulating his transformations. Lorentz considered that light originated from two sources, one stationary and the other in motion. The distance that light travels to reach an event from the moving frame needs correction factors to align the transformations with the results of the Michelson-Morley experiment and match the boat experiment. To achieve the principle of relativity in all frames of reference equally, Lorentz was compelled to add correction factors to the equation of the stationary frame, even though, according to Fitzgerald's principle, the distance that light travels does not change or contract because the frame from which the light beam originates is stationary and not facing the ether. Therefore, Lorentz's transformations become unrealistic and do not align with Fitzgerald's principle. Nevertheless, we will continue the derivation of Lorentz transformations.

Lorentz assumed that the moving frame started moving towards the event at the moment of coincidence with the stationary frame. At the same time, two light rays were emitted, one from the stationary frame and the other from the moving frame towards the event. It is natural for the light ray emitted from both frames to reach the event at the same moment because the speed of light is constant and is not dependent on the speed of the source, whether it is stationary or moving. This means that the times for the two light rays are equal.

It was necessary to determine the moment of measurement, which represents the moment of observation, and in turn, expresses the movement of the light beam. The observer closest to the event, who is in the moving frame, is assumed to observe the event before the stationary observer, who is considered to be farther away. However, the conditions for such observation were not available in Lorentz's transformations for both observers because the moment of observation is the result of the arrival of the light beam coming from the event to the observer's eye, and not the other way around.

Lorentz transformations were derived as shown below:

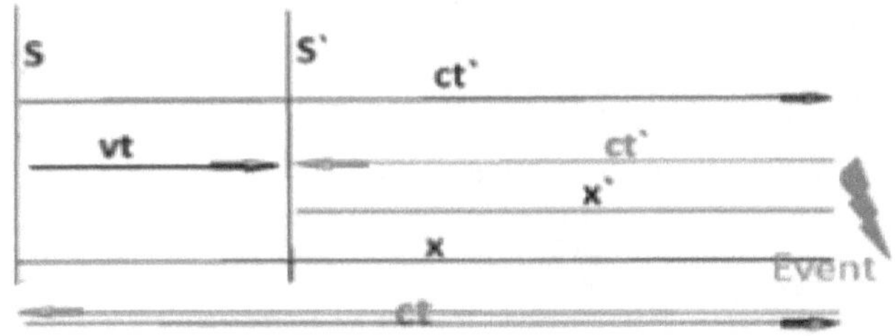

The derivation of Lorentz transformations resulted from multiplying the right-hand side of Galilean transformations by the length correction factor added by Lorentz to the equation. Then, by multiplying the two equations together and substituting the underlined coordinates in the equation below with the reciprocal of the speed of light, considering the speed of light is constant for all observers equally. The derivation of the equations is as follows:

$$x = \gamma\,(x' + vt') = ct \dots\dots (1)$$
$$x = ct \quad c = x/t$$
$$\underline{1/c} = t/x \dots\dots\dots(2)$$
$$x' = \gamma\,(x - vt) = ct' \dots\dots (3)$$
$$\mathbf{x' = ct'} \dots\dots\dots (4)$$
$$c = x'/t'$$
$$\underline{1/c} = \mathbf{t'/x'} \dots\dots\dots (5)$$

In equation (2), the time it takes for the light ray to travel from the stationary frame to the event is equal to the time it takes for the light ray to travel from the event to the observer in the stationary frame because they cover the same distance. Therefore, there is no difference in the times of the two light rays, and we can substitute one for the other.

The equation (4), which we highlighted in red, is indeed incorrect and led to an error in equation (5). The light ray that travelled the distance ct' in equation (3) moved from the moment of coincidence with the stationary frame until it reached the event. Since the motion of light is not dependent on the motion or speed of the source, the distance covered by the ray is not equal to X'. On the other hand, the ray that travelled the distance ct' moved from the event until it reached the moving frame after reaching S', so it covered the distance, X'. Therefore, the two rays cover different distances at various times, and one cannot be substituted for the other.

Where:

$$x' = ct'$$

$$x' \neq ct'$$

$$ct' \neq ct'$$

$$t' \neq t'$$

While Lorentz assumed a temporal coincidence, this is the critical error that made Lorentz transformations invalid, and all theories of special and general relativity built upon them are incorrect and need correction.

However, we will continue explaining the transformations to highlight the errors within them. By multiplying equations 1 and 3, we obtain the following equation:

$$xx' = \gamma^2 \left(xx' - x'.vt + x.vt' - v^2 tt' \right)$$
$$1 = \gamma^2 \left(1 - v \cdot \underline{t/x} + v \cdot \underline{t'/x'} - v^2 \cdot \underline{t/x} \cdot \underline{t'/x'} \right)$$

By substituting in the equation from the equations (2) and (5) with the reciprocal of the speed of light, the result becomes:

$$1 = \gamma^2 \left(1 - v \cdot 1/c + v.1/c - v^2 .1/c.1/c \right)$$
$$1 = \gamma^2 \left(1 - v^2 .1/c.1/c \right)$$
$$1 = \gamma^2 \left(1 - v^2 / c^2 \right)$$

Therefore, we obtain the Lorentz factor with the following mathematical formula:

$$\gamma = \frac{1}{\sqrt{1 - v^2/c^2}}$$

When you multiply the Lorentz factor by Galilean transformations, you obtain the equations that are supposed to represent the transformations between the stationary and moving frames, as shown below:

$$x = \frac{x' + vt'}{\sqrt{1 - v^2/c^2}}$$

$$x' = \frac{x - vt}{\sqrt{1 - v^2/c^2}}$$

These two equations are called the Lorentz transformations. For over 120 years, none of the brilliant scientists noticed or considered what these equations describe and their mathematical consequences. It is challenging to find a suitable description for them because the results of both equations do not represent anything we can define, and their numbers are imaginary and illogical, becoming more bizarre as the speed of the moving frame increases.

However, we can say that these two equations express that the observer in the stationary frame and the observer in the moving frame both see the event at the same moment, even if the observer in the moving frame has actually reached the event. This is the result of the matching of the time of the stationary frame's light ray and the time of the moving frame's light ray in the equation because they both cover the same distance. It should be noted that observing the event according to the equation is a result of the movement of the light ray emitted from the observer to the event, not the other way around, and this is an error that we need to correct. It was supposed that Lorentz transformations would achieve the principle of relativity and the transformation between different frames, but they completely failed to do so. Therefore, we find that Lorentz transformations do not match reality, do not replicate the Michelson-Morley experiment, do not comply with Fitzgerald's principle, do not suit the boats experiment, and do not even accurately describe the observation of events in their actual locations. Hence, they are strange transformations on which we cannot rely.

Afterward, Lorentz derived the equation of time, which represents the local time of the moving frame. It does not express time dilation but rather the spacetime interval between the two frames. This refers to the time it takes for light to travel the distance between the frames. You can convert between the times of the frames based on the distance between them. Time equals the distance divided by the speed, which is the speed of light, as follows:

$$t = \frac{t' + v\,x'/c^2}{1 - v^2/c^2}$$

$$t' = \frac{t - v\,x/c^2}{1 - v^2/c^2}$$

Below is a table of Galilean and Lorentz transformations for different velocities, illustrating the peculiar results using Lorentz transformations:

Time	Light	Dis	Vilocity	Trans.D	gama	Galilyeo	Galilyeo	constration	Lorenz x	Lorenz x´
t	C	d=x	vc	vt	$1/\sqrt{(1-v^2/c^2)}$	x´+vt´	x-vt	$\sqrt{(1-v^2/c^2)}$	x´+vt´/$\sqrt{(1-v^2/c^2)}$	x-vt/$\sqrt{(1-v^2/c^2)}$
t	C	d	V	D	γ	x	x´	L	x	x´=ct´
1	1	1	0,000001	0,000001	1,000000	1,000000	0,999999	1,000000	1,000000	0,999999
1	1	1	0,000928	0,000928	1,000000	1,000000	0,999072	1,000000	1,000000	0,999072
1	1	1	0,100000	0,100000	1,005038	1,000000	0,900000	0,994987	1,005038	0,904534
1	1	1	0,200000	0,200000	1,020621	1,000000	0,800000	0,979796	1,020621	0,816497
1	1	1	0,250000	0,250000	1,032796	1,000000	0,750000	0,968246	1,032796	0,774597
1	1	1	0,300000	0,300000	1,048285	1,000000	0,700000	0,953939	1,048285	0,733799
1	1	1	0,366025	0,366025	1,074570	1,000000	0,633975	0,930605	1,074570	0,681250
1	1	1	0,450000	0,450000	1,119785	1,000000	0,550000	0,893029	1,119785	0,615882
1	1	1	0,500000	0,500000	1,154701	1,000000	0,500000	0,866025	1,154701	0,577350
1	1	1	0,600000	0,600000	1,250000	1,000000	0,400000	0,800000	1,250000	0,500000
1	1	1	0,618034	0,618034	1,272020	1,000000	0,381966	0,786151	1,272020	0,485868
1	1	1	0,650000	0,650000	1,315903	1,000000	0,350000	0,759934	1,315903	0,460566
1	1	1	0,707107	0,707107	1,414214	1,000000	0,292893	0,707107	1,414214	0,414214
1	1	1	0,750000	0,750000	1,511858	1,000000	0,250000	0,661438	1,511858	0,377964
1	1	1	0,800000	0,800000	1,666667	1,000000	0,200000	0,600000	1,666667	0,333333
1	1	1	0,866025	0,866025	2,000000	1,000000	0,133975	0,500000	2,000000	0,267949
1	1	1	0,900000	0,900000	2,294157	1,000000	0,100000	0,435890	2,294157	0,229416
1	1	1	0,990000	0,990000	7,088812	1,000000	0,010000	0,141067	7,088812	0,070888
1	1	1	0,999900	0,999900	70,712446	1,000000	0,000100	0,014142	70,71245	0,00707
1	1	1	0,999990	0,999990	223,60736	1,000000	0,000010	0,004472	223,6074	0,00224
1	1	1	0,999999	0,999999	707,10696	1,000000	0,000001	0,001414	707,1070	0,00071
1	1	1	1,000000	1,00000	∞	1,0000	0,000000	0,000000	∞	0,000000

Lorentz's factor astonished scientists, both young and old, who considered it to rescue the transformations between reference frames and solve the problems of relativity. Among these scientists was the young Einstein, who was still in his twenties. He displayed more bravery than genius. Einstein applied Lorentz's factor to time, mass, velocity addition, and energy, giving birth to the theory known as Special Relativity.

Since Special Relativity failed to be applied to the motion of bodies under the influence of gravity, as they do not obey Newton's First Law, Einstein developed General Relativity for them. He spent ten years studying and deriving its equations and principles. Out of respect for the abilities and contributions of the great scientist Lorentz, no one dared to analyse, modify, or even interpret those famous transformations known as Lorentz transformations.

After Einstein published his Special Theory of Relativity, scientists ignored the transformations, as relativity itself became the centre of attention due to its uniqueness. We will delve into its study in depth and then analyse it with great precision.

Einstein's Train

Einstein was not to blame for deriving the equations of the Special Theory of Relativity in this non-realistic form. Lorentz transformations had fascinated him, and the equation for length contraction had occupied his mind. One evening, Einstein visited his friend, Michele Besso, hoping to find answers to the questions that had been bothering him and preoccupying his thoughts. As soon as Einstein shared his thoughts with Besso, a deep discussion ensued between them, lasting for hours, exploring all dimensions of the issue, delving into the finest details, and scrutinizing every aspect, but to no avail.

On the following day, as Einstein was commuting home by train and looking at the famous Bern Clock Tower, he asked himself the following question:

'If the train were moving away from the clock tower at the speed of light, would the clock appear motionless to him, while his personal wristwatch, unaffected by the train's motion, would continue to tick regularly?' This implies that events that are simultaneous in one frame of reference may not be simultaneous in another frame due to differences in the timing of events caused by their relative velocity. This is what Einstein envisioned.

Einstein believed that time runs at different rates in various places in the universe, depending on the speed of the body. The first ray of light coming from the clock tower at the moment the train departs will move at the same speed as the train, as they are both moving within the same frame, which is Earth. Therefore, the image of the clock that Einstein sees at the moment of departure is the same image, and what Einstein sees is nothing but an image coming from the past, while he moves with it at the same speed.

When Einstein got off the train, he was excited and uttered his famous words:

'A storm broke loose in my mind.'

He founded his theory of relativity, relying on his vivid imagination and leveraging Lorentz's transformations.

The Special Relativity Theory

When Einstein formulated the special theory of relativity, the time dilation equation was indeed counterintuitive compared to what one might expect with a stopped clock on a moving train. The time dilation equation suggests that time stops on the train when it moves at the speed of light, but it does not imply time stopping in a clock near the train station. In the equation, it is the image of the clock that moves with the train that stops, not the actual clock itself. The image of the clock remains stopped as long as the train continues to move at the speed of light, as Einstein envisioned.

It is impossible for light to be stationary, and it is the clock on the tower that moves away at the speed of light. At the moment the train stops, the image coming from the clock will start moving. This does not mean that time has actually stopped at the station because the clock on the tower will continue to move, whether the train is stationary or moving at the speed of light. There is no direct connection between the motion of the clock on the tower and the speed of the train. If the clock were the one moving at the speed of light and the train were stationary, its image would not appear to be stationary when observed by Einstein because the light coming from the clock is not related to the speed of the clock but rather to the distance the clock moves away. In this case, Einstein sees the clock moving slower than its natural motion because the beam coming from the clock will require more time the farther the clock moves.

Here, we see a case of a mismatch in the constancy of a physical law when it comes to the motion of light. The law that applies to a specific frame, does not apply to the other when the relationship is tied to the motion of light. While the principle of relativity requires the constancy of physical laws for all observers, regardless of their velocities. In other words, the result of any mathematical equation used by a moving observer must be similar to the result of the same equation used by a stationary observer. Therefore, this principle contradicts Einstein's theory of relativity.

If we assume that an object is moving at the speed of light according to Einstein's relativity equations, then the object becomes lengthless, the distance it covers becomes limitless, its mass becomes infinite, time becomes frozen, and it possesses immense energy. These properties are contradictory to each other since mass requires three dimensions, but length is absent. The object is moving at the speed of light and cannot traverse a distance of even one metre because time is frozen. The object's mass has become limitless, yet it does not exist as it has transformed into infinite energy.

When physicists began applying Einstein's relativity equations to some thought experiments, they encountered unreal results, far from logic and contrary to reality. Such as the famous twin paradox, time travel to the future, and many other mental experiments that defy all the laws of physics and contradict the laws of nature. They do not align with reality and are not in accordance with logic.

From the precise description of the relationship between the speed of light and the observer, and from the constancy of the laws of physics in all frames of reference, we can explain, discuss, and evaluate the validity of the special theory of relativity. Although the equations are derived from Lorentz transformations, which have been proven and confirmed to be incorrect, we search through the explanation to demonstrate their invalidity through other means, to remove the doubt and uncertainty that accompanies them.

Length Contraction Formula

$$L = L_0 \times \sqrt{1 - \frac{v^2}{c^2}}$$

This equation was formulated according to the principle of 'Fitzgerald', in the contraction of length in the direction of motion. The law expresses the difference in distance between two moving bodies in the same direction and at the same speed, in a frame of reference with inertia, as seen by an observer at rest with respect to the two bodies representing the ends of a single body moving away from the observer. The observer measures the distance between the ends of the body during motion by applying Lorentz transformations to each end of the moving body.

Albert Einstein had no role or contribution in the derivation of this law, as it was derived in collaboration between 'Lorentz and Fitzgerald'. The law was published in 1904 before Einstein published his theory of relativity, and this law is the first of the special theory of relativity.

The derivation of the law relies on measuring the difference in distance as perceived by the observer between each of the ends of the moving body, one close to the observer and the other far from them. To achieve this, the observation of both ends of the body must occur simultaneously. Since distance is equal to velocity multiplied by time, the time it takes to observe the far end of the body must be longer than the time it takes to observe the near end because it is farther away. As a result, the light ray coming from the far end takes a longer time. Despite this intuition, Lorentz considered that the time it takes to observe the near end to the observer is equal to the time it takes to observe the far end, as if the observation happened instantaneously with the light ray travelling from the observer to both ends of the moving body. In order for the observation of both ends of the body to occur simultaneously, the light ray coming from the far end must first reach the near end, and then the two rays progress together toward the observer, as illustrated in the diagramme below.

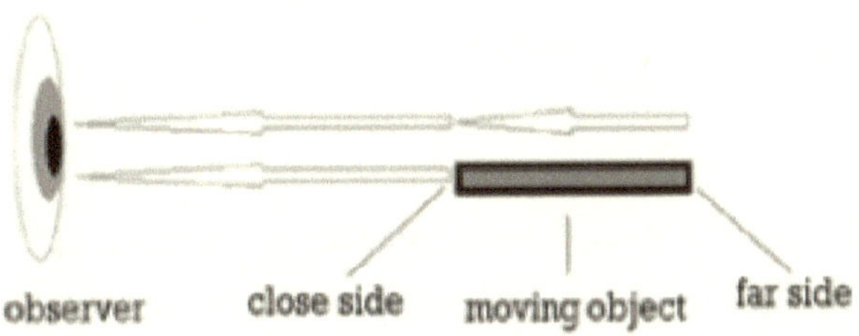

Lorentz's derive Law on Deflation of length, considering that both ends of the body, are bodies, and the distance between them is the length of the object to be measured in motion mode, without consideration, of the difference at the time of each party's viewing. The derivative is as follows:

$t_1 = t_2$

To determine the length of an object which is, in the event of stillness at the time of viewing, as in the illustration and equations below:

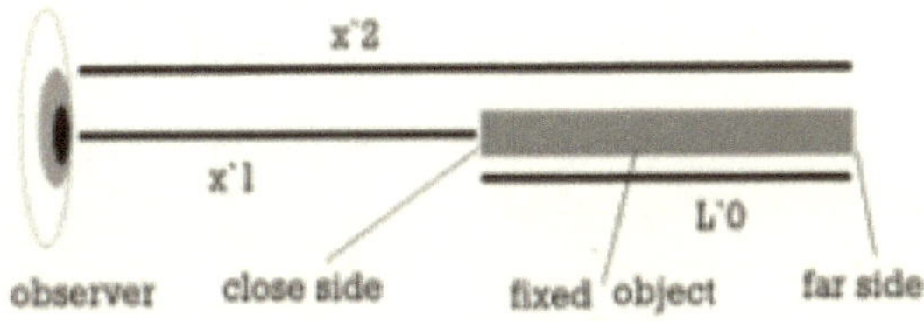

By linking the coordinates of the near end, which is at rest, to the coordinates of the observer, according to Lorentz's transformation.

$$x`_1 = \frac{x1 - vt1}{\sqrt{1 - v^2/c^2}} \quad \cdots\cdots\cdots\cdots (1)$$

linking the coordinates of the far end, which is at rest, to the coordinates of the observer, according to Lorentz's transformation.

$$x\grave{}_2 = \frac{x2 - vt2}{\sqrt{1 - v^2/c^2}} \quad \dots\dots\dots\dots (2)$$

Where:

X_1 - distance of the near end, relative to the observer.

X_2 - distance of the far end, relative to the observer.

$X\grave{}_1$ – distance of the near end at rest.

$X\grave{}_2$ – distance of the near end while moving.

t_1 – time it takes for the light coming from the near end to arrive.

t_2 – time it takes for the light ray coming from the far end to arrive.

V – speed of the moving object.

$L\grave{}_0$ – length of the object at rest.

L – length of the object as seen by the observer.

C – Speed of light.

object length during movement, for observer is:

$L = x_2 - x_1$

Then, its length during rest is:

$L\grave{}_0 = x\grave{}_2 - x\grave{}_1$

Since the times are equal, then the equation expressing the body's length, during rest, according to Lorentz's conversions, after subtracting equation (1) from equation (2). the equation becomes as following form:

$$L\grave{}_0 = \frac{x\grave{}2}{\sqrt{1 - v^2/c^2}} - \frac{x\grave{}1}{\sqrt{1 - v^2/c^2}}$$

$$L`_0 = \frac{x2 - v\,t2}{\sqrt{1 - v^2/c^2}} - \frac{x1 - v\,t1}{\sqrt{1 - v^2/c^2}}$$

$$L`_0 = \frac{x2 - x1}{\sqrt{1 - v^2/c^2}}$$

$$L`_0 = \frac{L}{\sqrt{1 - v^2/c^2}}$$

From the above equation, we obtain the length of the moving body as seen by the observer as follows:

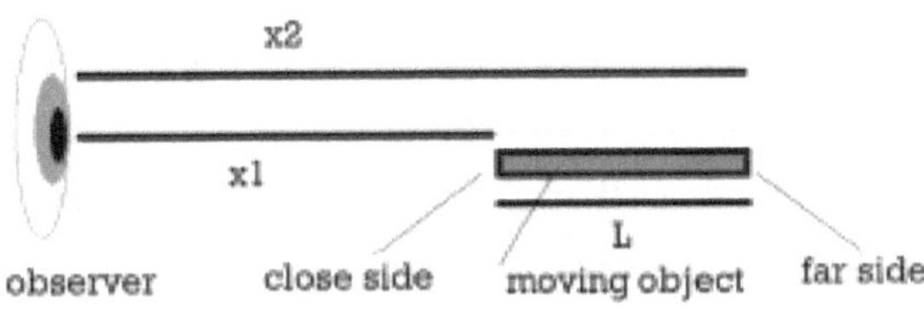

$$L = L`_0 . \sqrt{1 - v^2/c^2}$$

This is the mathematical form of the length contraction law, which is unrealistic because the time for observing the entire moving body is formulated in an unreasonable way. Even Lorentz did not consider the difference between observing the length of an approaching body and observing the length of a body moving away from the observer. The observation of the length of the body approaching should differ from the observation of the length of the body moving away. From the application of the equation to a body moving at the speed of light, the body becomes of zero length.

Furthermore, the equation expresses the following description: the distance between two bodies moving in the same direction at the speed of light has a value equal to zero, no matter how far apart they are. Does this mean that a body moving at the speed of light disappears from existence? And will it reappear if its speed decreases?

To answer this question, we need to find the correct equation that describes the actual length of the moving body when it is in a frame of reference with inertia, as observed by the observer. Let us assume that light travels a distance of one metre in one second, and we also assume that a body with a length of one metre is moving at the speed of light along a ruler. We can use this ruler to determine the position of the front of the moving body. The body starts moving from the beginning of the ruler at point zero. We have chosen its speed to be the speed of light to simplify the calculation process and explain the scenario. So, if we place an observer at the beginning of the ruler at point zero, the observer will see the front of the body at a distance of fifty centimetres after one second simply because the moving body's front becomes positioned at number 50 after half a second has passed. For the light beam coming from the front to reach the observer, it also needs half a second. During this half-second, the front has moved and covered a distance of fifty centimetres, placing it at number one hundred. At the same moment, the observer will see the rear of the moving body at the beginning of the ruler at number zero. Here, the condition of simultaneity in observing both the front and the rear at the same time is met. So, after one second, the observer will see that the length of the moving body is fifty centimetres, while in reality, its length is one metre. If we placed another observer at the end of the ruler at number one hundred, this second observer would see the front of the moving body after one second from the moment it started from the beginning of the ruler. This is because distance equals speed times time, and it represents the moment of observation for the first observer.

In this way, we find that the mathematical description of the experiment results in:

0,5 + 0,5 = 1 m

this result is in line with reality.

Increased Mass Formula

$$M = \frac{m_0}{\sqrt{\left(1 - \frac{v^2}{c^2}\right)}}$$

Every object in the universe is in motion relative to the absolute reference frame or is in motion relative to another frame of reference. This motion gives the object relative kinetic energy. The greater the absolute or relative velocity, the greater the kinetic energy. This kinetic energy can be transformed into thermal, mechanical, electromagnetic energy, or any type of energy, but it does not transform into mass. If we assume the transformation of this kinetic energy into mass according to the law of conservation of mass and energy as described by the equation, then the body loses its kinetic property, meaning it becomes stationary because it has lost its kinetic energy in favour of mass.

Therefore, we find that the equation describes the body as if it has two energies: one kinetic energy and the other that has transformed into mass. This description is illogical and scientifically unacceptable. Energy does not transform into mass in this way. Instead, it requires extremely precise and difficult conditions, such as tremendous pressure and elevated temperatures only found in the cores of stars, where temperatures are not less than ten billion Kelvin. The equation of mass-energy equivalence does not apply to kinetic energy but rather to radioactive elements that decay, transforming their mass into energy. The laws of physics do not change with changing velocity, as the law of mass used by the moving observer is the same law used by the stationary observer, and the result of the equation does not change with changes in velocity, or else the assumption of the constancy of physical laws becomes an unnecessary assumption.

Time Dilation

$$t = \frac{t`}{\sqrt{1 - \frac{v^2}{c^2}}}$$

The idea of the time dilation equation was based on the principle of the motion of a ball inside a train car when thrown vertically upwards. The ball falls vertically downwards in the same spot from which it was launched, from the perspective of an observer inside the train car. The ball's motion is influenced by two forces: the first is the velocity of the ball being thrown upwards, and the second is the horizontal velocity of the train. Therefore, the observer standing at the train station sees the ball moving with the resultant of these two velocities, perceiving its motion as curved rather than purely vertical.

The motion of the light beam matches the motion of the ball if there were a mirror on the ceiling of the train car reflecting light from the bottom to the top. However, Einstein made a mistake when he generalised the movement of light like this outside the train because light is not affected by the motion or velocity of the source outside the frame, while the motion of the ball continues to move with the resultant velocities even outside the train.

The diagram below illustrates the difference between the motion of the ball and the motion of light in a moving frame, as observed by the stationary observer at the train station when both the ball and the light exit the train car. If there were an opening in the ceiling of the train car for the light and the ball to exit, it is evident that the light would move independently of the train's motion and would not gain any kinetic energy. Meanwhile, the ball would have acquired the horizontal velocity of the train in addition to its vertical motion, as shown in the following diagram:

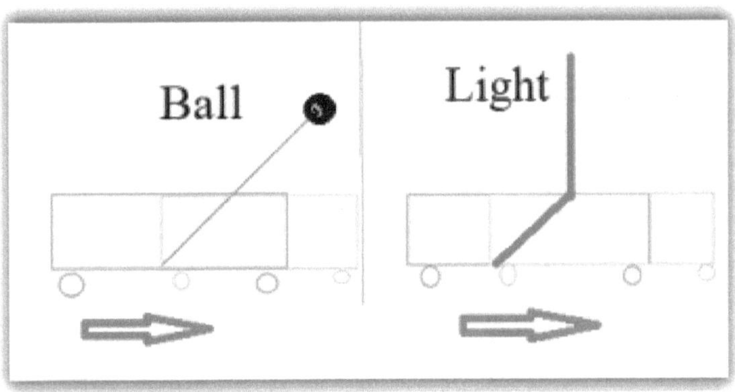

Based on this scientific fact, we must differentiate between the motion of light inside the frame and its motion outside the frame. Light moves at a constant speed only relative to the frame it is in because the frame is considered stationary with respect to the light. This is what the Michelson-Morley experiment confirmed.

The equation for time dilation, which is illogical, was derived, representing the triangle below, the coordinates of the distance travelled by the train, and the movement of the path of the light ray as seen by the observer inside the carriage, and the path of the light ray as seen by the observer standing at the train station, as follows:

Using a pulse of light that moves from the floor of the train carriage to the ceiling and reflects off a mirror on the train carriage's ceiling, the observer standing on the train station platform will see the light pulse in the passing train moving in a curved path rather than vertically, as observed by the observer inside the carriage. Since the curved path of the light is longer than the vertical path, it takes more time to travel from the ceiling of the carriage to the floor of the carriage for the observer on the platform. Given that the speed of light is constant, and the distance is fixed, the derivation of the equation using the Pythagorean Theorem, which represents the path of the light ray for the observers, and the speed of the train, is as follows:

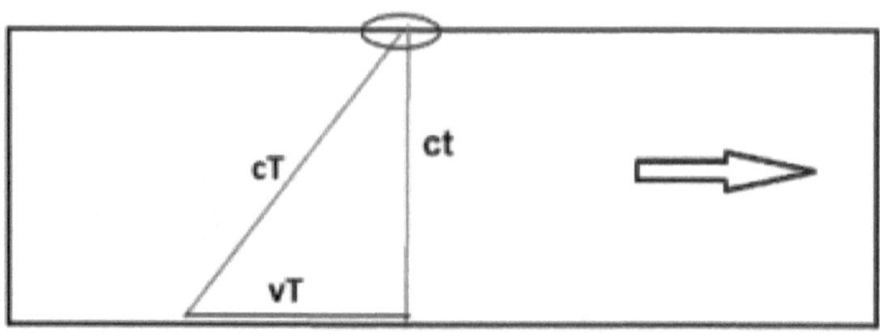

$$c^2T^2 = c^2t^2 + v^2T^2$$
$$T^2 = t^2 + v^2T^2/c^2$$
$$t^2 = T^2 - v^2T^2/c^2$$
$$t^2 = T^2(1 - v^2/c^2)$$
$$T^2 = t^2/(1 - v^2/c^2)$$
$$T = t \cdot \frac{1}{\sqrt{1 - v^2/c^2}}$$

Where:

C - Speed of light

V - Speed of the train

t - Time for the moving person

T - Time for the stationary person

In 1971, 66 years after the emergence of the special theory of relativity, when measurement devices and equipment became available, the astronomer "Keating" and the physicist "Hafele" wanted to conduct an experiment to verify the validity of time dilation. The experiment was as follows:

The two scientists prepared two airplanes and 12 HP5061A caesium beam atomic clocks.

These clocks were extremely precise, with almost no margin of error. They placed four of these clocks in each airplane and left four at rest on the ground at the United States Naval Observatory.

The first airplane flew eastward, in the direction of Earth's rotation.

The second airplane flew westward, against the Earth's rotation. The speeds of both airplanes were measured as follows:

Speed of the first airplane = Earth's rotational speed + Airplane speed.

Speed of the second airplane = Earth's rotational speed - Airplane speed.

Therefore, it was assumed that the speed of the first airplane would be faster, while the second airplane would be slower. As for the Earth's rotational speed around its own axis, it falls between the speeds of the two airplanes. Therefore, it is assumed that the clocks on the first airplane would be delayed compared to the other clocks because its airplane is faster. On the other hand, the clocks on the second airplane would be ahead of the other clocks because its airplane is slower. As for the clocks that remained on the ground, they are assumed to be ahead of the clocks on the first airplane and delayed compared to the clocks on the second airplane.

The ground speed of the airplane was 830 kilometres per hour.

The airplane's altitude is ten kilometres.

Therefore, the Earth's speed and the speeds of the two airplanes, as seen by an observer viewing the experiment from outside Earth's sphere, would be as follows:

The speed of First plane is:

$830 + 1674 = 2504$ km/h.

The speed of the second airplane is:

$830 - 1674 = 844$ km/h.

The Earth's rotational speed around its axis is

1674 kilometres per hour.

At the end of the experiment, both airplanes landed after orbiting the Earth's sphere. However, the experiment did not yield any satisfactory results, and its outcomes did not support the validity of the time dilation equation. In fact, some of the airplane clocks were ahead of the others, not delayed. To cover this failure, "Keating" claimed that the experiment confirmed the accuracy of time dilation near massive objects, as stated in the General Theory of Relativity.

It is ironic to accept the validity of the experiment's results and their alignment with either the Special or General Theory of Relativity. We can say that the Earth moves at a speed of 30 kilometres per second in its orbital motion around the Sun, but the Earth's rotation around itself does not result in any linear motion, even if the Earth were rotating at the speed of light. Therefore, adding and combining the speed of the two airplanes with the Earth's rotational speed is far from logical. The essence of motion is to cover a certain distance in a certain time, not rotating in a fixed place.

Simply put, we can discover the error in this experiment because it does not apply to the theory of time dilation. The two airplanes violated a crucial condition of the Special Theory of Relativity, which is motion according to the principle of inertia. The airplanes were subjected to external and self-induced forces, did not move in straight lines, and returned to their departure points. More so, the airplanes were flying within Earth's frame of reference. Therefore, adding or subtracting the airplane's speed from Earth's speed is an incorrect procedure because it neither increases nor decreases the airplane's speed. For the airplanes, the Earth is considered a stationary frame. So, if we assume that the airplane's speed is equal to the Earth's rotational speed, does that mean the airplane flying against the direction of Earth's rotation would be at rest with a speed of zero? And how would you measure its speed in that case?

We have a practical real-world experiment based on the GPS satellite system. These satellites travel at a speed of 14,000 kilometres per hour at an altitude of 20,200 kilometres above sea level. The operation of these satellites relies on the time it takes for signals to travel between the satellite and the ground station. Any difference in timing between them results in an error in determining the location on Earth. Therefore, the time difference in the clocks between the ground station and the satellite clocks is corrected several times daily. Physicists have divided into two teams in explaining the reason for the time difference between Earth and GPS satellites. The first team claims that time slows down in GPS satellites because they move at a high-speed relative to Earth, as described in the theory of special relativity. Therefore, the clocks on GPS satellites are adjusted based on this principle. The second team of physicists argues that time in the clocks of GPS satellites accelerates relative to the clocks on the Earth station due to their distance from Earth's massive body, according to the theory of general relativity. Hence, the clocks on GPS satellites are adjusted because they run faster relative to the clocks on the Earth station.

The reader has the option to interpret and analyse what happens to the clocks on GPS satellites. Are they slowing down or speeding up according to Einstein's theories? Or are there other reasons that lead to the clocks on GPS satellites experiencing either deceleration or acceleration? These reasons could potentially include variations in temperature, differences in atmospheric pressure, or even the absence of air in outer space, which may cause changes in clock speed. Alternatively, there could be other factors at play, such as the time delay in the electromagnetic wave transmission between satellites, ground stations, and mobile phones.

Since the satellite receives the signal and retransmits it to the ground station, what is the relationship between the signal and the timing of the clocks on the satellites? And why aren't the ground clocks adjusted? Doesn't the same law apply to both the Earth and the satellites?

There is a real paradox posed by Professor Stephen Hawking. Stephen Hawking, the theoretical physicist at the University of Cambridge, organised a large special event for time travellers on June 28, 2009. It was as follows:

And on the following day, which is the 29th, the invitation for time travellers was published in the local newspapers. The invitation, hosted by Professor Stephen Hawking, was for an event at the University of Cambridge, specifically at the Hall of the Faculty of Science, located on Trinity Street in Cambridge, at the coordinates:

52°12'21" N, 0°07'4.7" E

The event was scheduled for 12:00 PM on June 28, 2009. Stephen Hawking waited eagerly on that day, but no one attended.

This event served as a message from him that all we hear about time travel is science fiction and cinematic scenarios.

As Stephen Hawking mentioned in his book, "Brief Answers to the Big Questions":

"If someone were to apply for a grant to study time travel, they would be immediately dismissed, according to the American website, Science Alert. This is because the question of the possibility of time travel is an extremely serious one and cannot be addressed scientifically because any research that proves the impossibility of time travel is, in reality, a refutation of the theory of relativity, one of the most important principles of modern physics."

Einstein relied on the constancy of the speed of light for all observers equally when deriving his theory. Therefore, time must be the only variable in the equation, as long as distances do not contract as "Einstein" believed. Time is constant and flows uniformly throughout the universe from the moment of the universe's inception. What observers perceive as a difference is the temporal dimension between the two frames, and it is not a difference in or slowing down of time.

The temporal dimension between two frames can be found using the following equation:

The distance travelled by the moving object relative to the stationary observer is equal to the distance travelled by the light coming from the moving object to the observer. From this, we can find that the time difference between two locations is equal to the distance between them divided by the speed of light.

$cT=vt$

$T=vt/c$

Where:

T - Time at the location where the object arrives as seen by the observer.

t - Time at the observer's location.

c - Speed of light.

v - Speed of the moving object.

The speed of light vertically within the moving frame, as seen by an observer in the stationary frame and considering it as a second frame, can be described as follows:

t - The time it takes for the object to travel a certain distance.

cf - The speed of light within the moving frame relative to the stationary observer.

$$(c_f t)^2 = (ct)^2 + (vt)^2$$

$$c_f^2 t^2 = c^2 t^2 + v^2 t^2$$

$$c_f^2 = c^2 + v^2$$

$$C_f = \sqrt{c^2 + v^2}$$

This is the same law, with respect to observers in a moving frame who understand something clearly at last emitted from an object in a stationary frame. The speed of light differs for the observer when the motion of light moves from one frame to another. Therefore, we find that the speed of light within a moving frame, as seen by an observer in a stationary frame, is the sum of the speed of light added to the speed of the frame in which the light is moving when the motion is in the same direction. When the light is moving in the opposite direction to the motion inside the frame that is moving away from the stationary observer, the speed of light is subtracted from it.

According to the concept of relativity, to describe the motion of a vertical light ray inside a train carriage as seen by an observer standing on the platform, we can describe it as follows: If we assume that the train is moving at the speed of light, then the vertical light ray's length would be equal to the distance the train covers in the same time period. This is the maximum speed the train can move, as illustrated in the graph below:

$$C_t = V_t$$

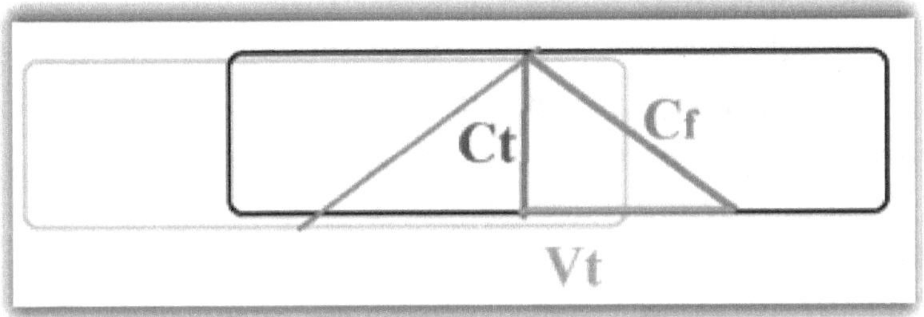

If the length of the vertical ray is one metre, and the train is moving at the speed of light, then the distance the train covers in the same time period is also one metre. Therefore, the maximum length of the hypotenuse representing the light ray that the stationary observer is supposed to see is:

$$C_f^2 = ct^2 + vt^2$$

$$C_f^2 = 1^2 + 1^2$$

$$C_f = \sqrt{2}$$

$$C_f = 1{,}41421356237331 \text{ m}$$

This is the maximum length for the hypotenuse, and it cannot be longer than that unless the train is travelling at a speed exceeding the speed of light.

Therefore, we find that the time dilation equation does not match the practical experiment, and it contradicts itself. In addition, we cannot disregard the length contraction in the experiment, where the length of the hypotenuse becomes equal to the length of the vertical line due to length contraction to the same extent caused by the difference in length on the expected time dilation. In other words, the time for the light ray seen by the observer inside the train becomes equal to the time for the ray seen by the stationary observer standing at the train station. We explain this below:

According to Einstein's concept of relativity, to understand the effect of speed on the physical properties of a moving object, we must apply all the equations of special relativity together, and it is not permissible to apply one equation and neglect the others. When we do that, we find that the physical properties remain constant and do not change with changes in speed. We can easily confirm this by conducting a mental experiment with a train car, one metre in length, emitting a flash of light from its rear, heading towards the front of the car. It travels one metre in one second when the train is stationary at the station, assuming that light travels one metre in one second. If the car is moving at half the speed of light, what will its length be, and how will its time dilate? When applying the length contraction equation first:

$$L = L_0 / \sqrt{\dfrac{1}{1 - v^2/c^2}}$$

$$L = 1 * \sqrt{0{,}75}$$

$$L = 0{,}866 \text{ m}$$

Where:

L – The length of the train car during motion.

L0 – The length of the train carriage at rest.

V – The speed of the train.

C – The speed of light inside the train carriage.

This means the distance actually contracted from one metre to 0.866 metres.

According to the relativity equations, time dilation in the carriage when speed is half the speed of light is:

$$T = t. \frac{1}{\sqrt{1 - v^2/c^2}}$$

$T = 1/0,866$

$T = 1,1547344$ Sec

Which is the time that light needs to travel the entire length of the carriage in its moving state, without contraction.

To calculate the time required to cover the distance of the contracted length of 0.866 metres, we multiply the distance by the time for the train carriage as follows:

$1,1547344 * 0,866 = 1$ Sec

The higher the relative velocity, the more it causes a contraction in the distance travelled, making the time experienced by the moving observer match the time of the stationary observer.

If we applied the time dilation equation first, we would find that the length would not change with a change in velocity.

The Wonder of Velocity Addition Law

$$V =$$

$$\frac{u1+u2}{1+(u1.u2)/c^2}$$

Einstein derived this equation by combining two equations from Lorentz transformations, after dividing both equations by time. The final equation represents the sum of the velocities of the two frames, where the sum cannot exceed the speed of light, as shown below:

$$V = \frac{\Delta x}{\Delta t}$$

$$u = \frac{x}{t} = \frac{(x'+ut')\gamma}{\left(t'+\frac{ux'}{c^2}\right)\gamma}$$

$$V = \frac{\frac{(x'+u1t')\gamma}{\left(t'+\frac{u1x'}{c^2}\right)\gamma} + \frac{(x'+u2t')\gamma}{\left(t'+\frac{u2x'}{c^2}\right)\gamma}}{}$$

$$V = \frac{u1+u2}{1+(u1.u2)/c^2}$$

Where:

V - Sum of the velocities

U1 - Velocity of the first body

U2 - Velocity of the second body

x - Distance

t - Time

c - Speed of light

γ - Lorentz factor

What is strange in relativity is the law of addition of velocities, where the equation describes that every object in this universe is absolutely at rest when subtracting or adding the velocity of any object to the velocity of light. The result becomes the speed of light. Since light travels at this constant speed, it means that the object is stationary no matter what its speed is.

We can easily find the flaw in this law. When the relative velocity between two objects is the speed of light, if we consider one of them to be at rest and the other to be moving at the speed of light, then the sum of their velocities is the speed of light. However, if we consider both objects to be moving at half the speed of light, the sum of their velocities would be 0.8 times the speed of light. In this case, no one can determine which object is faster or if either of them is stationary. So, which result is correct? When it comes to the speed of light itself, we cannot apply the principle of the constancy of physical laws because light must always travel at its constant speed, and the other frame must be the stationary one. In this case, the relative velocity between the two frames becomes the speed of light, and it cannot increase or decrease regardless of the speed of the moving frame. Here, the formula of the law that fits the motion of the speed of light is revealed, not the speed of objects. It is not permissible to substitute the motion of light with the motion of a moving body because light is independent in its direction and speed from the direction and speed of the frame. The issue becomes more complex when you add or subtract the velocities of two beams, as the result will always be the speed of light, and neither of them can be considered stationary or moving.

The equation for adding velocities can represent what an observer on Earth sees when looking at a laser beam emitted from an airplane and measures its speed. However, it doesn't represent the sum of two velocities because the observer is only measuring the speed of light. The speed of light is not dependent on the speed of the source.

The flaw in this law becomes evident when adding the speed of a fast boat that emits a laser beam in the water towards a submarine in front of it. In this case, according to this law, the speed of the laser beam becomes faster than the speed of light in the water. This demonstrates the failure of the law and the inaccuracy of its results. The reason is that the speed of light in the equation is not a universal constant but rather the speed of light in the medium described by the situation. The same issue arises for those working in nuclear reactors when observing the speed of the Cherenkov radiation, which is produced by enriched uranium submerged in heavy water. This radiation is named after its discoverer, the Russian engineer "Cherenkov". When monitoring and measuring the speed of Cherenkov radiation using the equation for adding speeds, Cherenkov radiation is faster than light. Upon closer examination, it becomes apparent that the speed of light in the equation is not a universal constant but rather the speed of light in the frame in which the light is moving. This applies to all equations in the theory of special and general relativity. Therefore, we find that the speed of light in water is faster than what is expected in the results of the equation for adding speeds.

The equation also does not apply to a frame moving in one direction with a light beam emitted in the opposite direction. In this case, the final speed must be the sum of the two speeds because light will travel at its fixed speed independently of the frame's speed.

The Law of Energy

$$E=mc^2$$

The relationship between mass and energy was not a new idea; it had been discovered by other scientists long before Einstein.

The first person to discover the equation for energy was the scientist Gottfried Leibniz, and he called it vis viva or self-energy. He mentioned this equation in a letter he wrote to the French physicist Emile du Châtelet, who pointed out a flaw in Newton's second law. Du Châtelet wrote several letters to various scientists, asking for their opinions on the matter. Leibniz's response aligned perfectly with the viewpoint of Marquise du Châtelet. Emilie du Châtelet eventually found her answer in the Dutch scientist Willems Gravesande, who conducted an experiment that confirmed the validity of the equation, according to Du Châtelet's physics, by dropping lead balls from different heights and finding that energy was proportional to the square of velocity, not double.

According to the law of conservation of energy, any object that possesses kinetic energy, potential energy, or any other form of energy will retain that energy until it is transferred to another object, depending on the type of energy it possesses. This includes potential energy, which represents the energy of an object when it is at rest and depends on the position or state of the object in relation to other objects.

Kinetic energy resulting from velocity:

$KE = 1/2mv^2$

Where:

KE - Kinetic energy

m - mass

V- velocity

When an object moves at the speed of light, its kinetic energy becomes as follows:

$$E = 1/2 \,.mc^2$$

The equation represents the energy of a stationary object, where its mass is transformed into energy. This energy, resulting from mass loss, is synonymous with the energy produced when atoms of the same mass combine, with the collision speed of the merging atoms being the speed of light.

As for Einstein's relativistic energy equation, it was derived by multiplying the kinetic energy equation by the Lorentz factor. This equation represents the energy of a static object, making it non-kinetic energy but energy resulting from mass conversion. It can be expressed as follows:

$$KE = (1/2mv^2) * \frac{1}{\sqrt{1 - V^2/C^2}}$$

Accordingly, the equation becomes as follows:

$$E = mc^2$$

The physicist Emilie du Châtelet mentioned the equation for self-energy in her book: "Institutions de physique", which she published in 1740.

INSTITUTIONS

DE

PHYSIQUE.

A PARIS,
Chez PRAULT fils, Quai de Conty, vis-à-vis la
descente du Pont-Neuf, à la Charité.

M. DCC. XL.
Avec Approbation & Privilège du Roi.

Einstein used this equation in his postulates in the theory of special relativity, and he multiplied it by the Lorentz factor, making it the most famous equation in the world, known as the equivalence of mass and energy, as if energy and mass are two sides of the same coin. The mass of radioactive elements can be converted into energy, but this does not apply to the rest of the elements. In order for energy to be converted into mass, this can only happen theoretically in mathematical equations, and it is impossible, except in impossible conditions, because it requires enormous pressure and temperatures of up to ten billion Kelvin, which are only available in the core of distant stars. Therefore, we can either say wave energy or we can say the mass of an object, and we cannot refer to an object as both mass and wave energy at the same time. As for the amount of energy produced by the transformation of matter into energy, credit for this goes to the physicist Lisa Meitner.

On a chilly day in the winter of 1907, the 28-year-old Austrian physicist Lisa Meitner, who held a Ph.D. in physics from the University of Vienna, boarded a train from Vienna to Berlin. She was determined to find work in the field of radioactivity. Unfortunately, German universities did not hire women as professors, so she was appointed as an overseer in the wood storage area of the chemistry institute. Otto Hahn, a chemist at the institute, stood by her side, Max Planck advised him to collaborate with Meitner due to the interconnectedness of chemistry and physics in radioactivity research. Hahn established a laboratory for this purpose.

At that time, extraordinarily little was known about the atom, except those electrons revolved around its nucleus. However, radioactive elements like radium and uranium were unstable even in their nuclei, emitting energy and particles from their mass, as if the mass of their nuclei were transforming into energy.

The collaboration between Otto Hahn and Lisa Meitner in nuclear fission began in 1911 at the newly established Kaiser Wilhelm Institute for Chemistry, later named the Max Planck Institute after World War II. Lisa Meitner became the first woman in Germany to obtain the title of professor at German universities.

One of the most famous elements they worked with was uranium, a recently discovered and very heavy element compared to iron. The nucleus of a uranium atom contains 238 protons and neutrons, and each atom has ninety-two electrons orbiting around the nucleus. The goal was to bombard the uranium nucleus with protons to create a larger nucleus, and from it, obtain a new substance.

When the Nazis came to power in the early 1930s, Germany became a dangerous place even for scientists. Foreign academics were expelled from universities, and Lisa Meitner remained under surveillance until 1938. She was eventually expelled from the university after the Nazis annexed Austria.

Hahn struggled to protect her, but she had no passport to leave Germany, and she was prohibited from leaving Germany when physicists from various parts of Europe invited her to attend conferences outside of Germany. With the help of one of her colleagues, she was smuggled to the Netherlands, from where she travelled to her sister in Sweden.

Meitner lost everything: her job, her home, and all her possessions. She had spent over thirty years in the service of science, yet she continued to communicate with Hahn through letters.

On one Christmas day, in the presence of her nephew Otto Albert Frisch, who was also a physicist, she received a letter from Hahn. In the letter, Hahn informed her that a few months ago, the strange thing he had observed when bombarding uranium atoms with protons was not an increase in the size of the atom, as initially believed. Instead, he detected what seemed to be contamination with radium, a much smaller element than uranium. He also noticed the presence of barium atoms, which are much smaller than uranium atoms, in his recent experiments. Hahn was sure that there was no error in the experiments and needed Meitner's expertise to make sense of this discovery.

Scientists used to believe that bombarding the nucleus of an atom with protons would increase its size and weight. However, Lisa Meitner, with the help of her nephew Otto, realised that when an atom grew too large, it might become unstable and split into two parts. This split would release a tremendous amount of energy. By calculating the energy released from the collision of two nuclei, Meitner was able to estimate the energy produced from the fission that resulted in two barium atoms, with a value of two hundred million electron volts. But where did this energy come from?

Through some calculations, Meitner found that the masses resulting from the fission were five protons less than the nucleus of the original uranium atom, meaning there was a missing mass. From the energy equation, she found that the energy released from the missing mass was equal to the energy needed to fuse two nuclei of barium atoms. This was a remarkable discovery. She had calculated the energy produced from nuclear fission. Lisa Meitner published this discovery, which she called "nuclear fission".

In 1944, Otto Hahn received the Nobel Prize after claiming that he had discovered nuclear fission, without mentioning Meitner's contribution. This deeply saddened Meitner, as she became overshadowed by Hahn's previous assistant.

Physicists around the world became aware of this discovery, and in the United States, the Manhattan Project, a secret initiative to build the atomic bomb that ended World War II, began as a result.

If it were possible to convert energy into mass, sound would also have mass. The energy of a sound wave is significant, and it manifests itself in the phenomenon of sonic booms, which we can describe as follows:

Before we see lightning, we hear the sound of thunder that it produces. Every explosion, whether it is electrical, from a fire, or a nuclear blast, generates a sound wave.

It turns out that the matter that converts into energy transforms into rectangular waves that produce sound and into exposed waves that appear as light. Therefore, sound has energy, just like light has energy. If photons of light have mass according to the mass-energy equivalence equation, then sound also has mass. This energy appears in the phenomenon of acoustic luminescence that occurs for a sound wave in a bubble of air underwater. However, plants can easily and remarkably convert electromagnetic wave energy into mass by interacting with carbon dioxide gas.

Below is a table of different measurements in Einstein's special relativistic theory equations.

Time	Light	Distance	Length	Mass	Velocity	Velocity 2	Relative V	Lorentz f	length contraction	Mass	time dilation	Acceleration	Energy
sec	C	D=X	L	Kg	vc	Vc2	vc	γ	d	M	T	U	E
t	(m/s)c	m	m	M	m/s	m/s 2	m/s	$1/\sqrt{1-v^2/c^2}$	m	Kg	s	m/s	jc
1	1	1	1	1	0,00001	0,00001	1	1,00000	1,00000	1,00000	1,00000	0,00001	1,00000
1	1	1	1	1	0,01000	1,00000	1	0,99995	0,99995	1,00005	0,99995	1,00000	1,00000
1	1	1	1	1	0,10000	0,90000	1	0,99499	0,99499	1,00504	0,99499	0,91748	1,00000
1	1	1	1	1	0,20000	0,80000	1	0,97980	0,97980	1,02062	0,97980	0,86107	1,00000
1	1	1	1	1	0,25000	0,75000	1	0,96825	0,96825	1,03280	0,96825	0,84111	1,00000
1	1	1	1	1	0,30000	0,70000	1	0,95394	0,95394	1,04828	0,95394	0,81646	1,00000
1	1	1	1	1	0,36603	0,60000	1	0,93060	0,93060	1,07457	0,93060	0,79107	1,00000
1	1	1	1	1	0,45000	0,50000	1	0,89303	0,89303	1,11979	0,89303	0,77551	1,00000
1	1	1	1	1	0,50000	0,40000	1	0,86603	0,86603	1,15470	0,86603	0,75000	1,00000
1	1	1	1	1	0,60000	0,20000	1	0,80000	0,80000	1,25000	0,80000	0,71429	1,00000
1	1	1	1	1	0,61803	0,10000	1	0,78615	0,78615	1,27202	0,78615	0,67634	1,00000
1	1	1	1	1	0,65000	0,01000	1	0,75993	0,75993	1,31590	0,75993	0,65574	1,00000
1	1	1	1	1	0,70711	0,00000	1	0,70711	0,70711	1,41421	0,70711	0,70711	1,00000
1	1	1	1	1	0,75000	0,10000	1	0,66144	0,66144	1,51186	0,66144	0,79070	1,00000
1	1	1	1	1	0,80000	0,20000	1	0,60000	0,60000	1,66667	0,60000	0,86107	1,00000
1	1	1	1	1	0,86603	0,30000	1	0,50000	0,50000	2,00000	0,50000	0,92556	1,00000
1	1	1	1	1	0,90000	4,00000	1	0,43589	0,43589	2,29426	0,43589	1,06511	1,00000
1	1	1	1	1	0,95000	0,50000	1	0,31225	0,31225	3,20256	0,31225	0,98305	1,00000
1	1	1	1	1	0,99900	0,60000	1	0,04471	0,04471	22,36627	0,04471	0,99975	1,00000
1	1	1	1	1	0,99990	0,70000	1	0,01414	0,01414	70,71245	0,01414	0,99998	1,00000
1	1	1	1	1	0,99999	0,80000	1	0,00447	0,00447	223,60736	0,00447	1,00000	1,00000
1	1	1	1	1	1,00000	1,00000	1	0,00000	0,00000	#DIV/0!	0,00000	1,00000	1,00000

Chapter III

Relativity in Artificial Intelligence

Doppler Effect

It is known that light is affected by the Doppler Effect, which was discovered by the Austrian scientist Christian Doppler in 1842. Doppler found that the time required for one wave cycle, which is the period, is equal to the reciprocal of the frequency, which is equal to the speed of the wave divided by its wavelength, as expressed in the following equations:

$T = 1/F$

$F = c/L$

The Doppler Effect contradicts Einstein's interpretation when he considered the speed of light to be constant for all observers in different frames of reference equally. This is because the speed of the frame does not change its relative speed with the light it emits; instead, it changes the wavelength. The change in wavelength implies a difference in the distance that the wave moves away from the frame it originated from, whether in the same direction or in the opposite direction. This means that the relative distance covered by a light wave in the direction of motion of the moving frame is less than the relative distance covered by a light wave emitted from the stationary frame. Consequently, if an object approaches an observer at the speed of light, the observer will not be able to see the object before it reaches them. In other words, the light emitted by the object cannot precede the object itself, as the light ray has effectively covered a distance of zero relative to the frame it originated from, making the wavelength of light effectively zero. This is analogous to not hearing the sonic boom of a supersonic jet until it has already passed us.

To determine the speed of light relative to an observer in a moving frame, a beam of light is emitted forward from the frame in the same direction as its motion. The speed can be found by calculating the difference between the wavelength of the emitted beam when the object is at rest and the wavelength of the beam outside the moving frame during motion.

The following figure illustrates the distance between the frame and the light pulse, and the difference in the distance covered by the light emitted from two frames, one stationary and the other in motion.

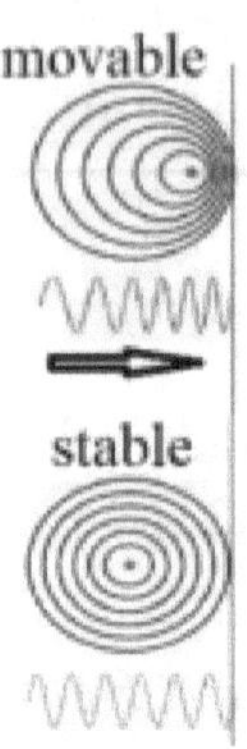

In the Michelson-Morley experiment, there was no Doppler effect observed on light because the light source, detector, and observer were all within the same frame of reference, the Earth, which is considered to be stationary.

Maxwell's equations showed that electromagnetic waves propagate in a vacuum at a constant speed relative to a stationary reference frame in the vacuum, which scientists originally called "the ether". Scientists believed that the ether was the material and medium through which light travelled, but later discoveries revealed that space fills the universe. Even within the tiniest particles of matter, such as the hydrogen atom, more than 99.999% of the space is empty. Therefore, electromagnetic waves propagate in a vacuum under all conditions, in all places, and through all materials, whether it is air, glass, or any other substance.

To put it in perspective, if the size of a hydrogen atom were the size of a football field, then the nucleus of the atom, the size of a tennis ball, would be located at the centre of the field. The electron, which might be the size of a lentil, would orbit outside the bleachers of the field. The rest of the field is entirely empty space through which light travels freely.

Scientists have discovered that most of the universe is vast emptiness between galaxies, containing no matter, with these voids spanning billions of light-years in diameter. Therefore, the actual meaning of the "Ether" is the vacuum, representing absolute stillness through which light travels, relative to it. The vacuum within the particles of matter follows their motion relative to the frame carried by the particles. This makes the speed of light constant in all reference frames for all observers, each within their own frame, but not equally. Every frame in the universe moves relative to a reference frame of absolute stillness, which can be called "the ether", not as a material medium that carries light waves, but as a reference frame of absolute stillness alone. Light moves in the vacuum at its constant speed relative to it, regardless of the frame from which it originates.

The Michelson-Morley experiment did not match the boat experiment, but Earth's motion in the stationary ether matches the movement of a ship representing Earth, which moves in the calm sea representing the ether. The movement of any object within the ship is relative to the ship, and even light, inside the ship and even on the surface of the ship, moves at its constant speed in all directions relative to the ship. However, the movement of any object, including light, when it exits the ship into the sea, becomes relative to the stationary sea and is not related to the speed of the ship. Here, we find a difference in the speed of light relative to the ship when light transitions from the ship's frame to the frame of the sea.

In the Michelson-Morley experiment, the light beam was still moving under the influence of Earth's motion and speed, and it did not exit into the ether.

Although light requires a vacuum to propagate through, it is influenced by the motion of matter, the transmitting medium, and the gravitational field. As light moves farther from Earth, the Earth's medium has less impact on it, while the medium influenced by the Sun has a greater effect. The point at which the influence of both the Earth and the Sun equals out is estimated to be more than 1.5 million kilometres away from Earth. The same effect also occurs on a spacecraft travelling to Mars; it orbits with the Earth's rotation. When the Earth's influence diminishes, the spacecraft is freed from Earth's rotation and starts orbiting with the solar space until it reaches Mars.

The Earth's influence does not appear to affect the movement of light coming to it due to its immense speed, but we can observe this effect when light from distant stars located behind the Sun passes through it. The light is affected by the motion, speed, density, temperature, and other properties of the solar-transmitting medium, causing stars to appear not to be in their correct positions. This effect is not due to gravity but rather the influence of the transmitting medium on its direction, velocity, density, temperature, and other properties.

We can conduct an experiment to demonstrate the effect of the speed and direction of the transmitting medium on the speed and direction of light, confirming that it is not due to gravity. This can be done by passing light through a rapidly rotating glass disc and observing the light beam as it passes through the disc towards a background behind it. There will be a noticeable shift in the direction of the light beam after the disc is rotated at an extremely high speed. This experiment confirms the influence of the motion of the transmitting medium, which is not the effect of gravity.

There is an optical experiment conducted by the French scientist "Fizeau" in 1851, where he used a special system to measure the effect of a moving medium on the speed and direction of light. He found that the speed of light in water is less than in air, and that moving water drags the light along with it in the direction of its motion. These pieces of evidence are sufficient to prove that light moves in accordance with the motion of the frame to which it belongs.

The speed of a sound wave increases as the density of the medium it travels through increases, while the speed of an electromagnetic wave decreases as the density of the medium increases. Therefore, we find that the speed of light is at its highest in a vacuum when there is no matter present, and it is approximately 299,792,458 metres per second. In contrast, the speed of light in the air is slightly less than that, around 299,702,547 metres per second. The speed of light in water is approximately 225,000,000 metres per second, and it loses some of its energy when it reflects off the surface of the water. It continues to lose energy gradually in the water, transforming into thermal energy, heating the water until it eventually dissipates in the depths without losing any of its speed.

It has become entirely clear that the symbol 'C', which represents the speed of light, is not a universal constant, as claimed by Einstein, because it represents the speed of light only in a vacuum, not in air, water, or any other substance. Its speed is constant only relative to the observer in the same frame of reference. This description changes the results of all equations in which the speed of light is considered a universal constant, including all of Einstein's theories of relativity. The speed of light is constant relative to the medium through which it travels, and the speed of light varies with the relative speed between different frames of reference, each according to its direction and relative speed concerning the frame in which the light is moving.

That emphasises the need to reconsider the Lorentz transformations because they indicate that the observer in the stationary frame and the observer in the moving frame see the event at the same time. However, we are all certain that the observation of the event happens differently and depends on the distance between each observer and the event, unrelated to the speed of the observer. Therefore, we have derived transformations that take into account the times of different events among all frames of reference, located at different distances from the event and at various times. These new transformations are to be applied to all physical laws, including Maxwell's equations. This means that the nearby observer sees the event before the distant observer. All we will do is derive transformations by applying Galilean transformations to the beam emitted from the event to the observers, instead of the beam that Lorentz derived, which was in the wrong direction.

New Transformations

To explain the derivation of transformations that express the real distance between the first observer, whom we consider stationary, the second observer, whom we describe as moving, and the event. We must remember that observation is achieved the moment the light emanating from the event reaches each observer, according to his location and the distance between him and the event, at the moment of measurement. Regardless of his relative speed, it does not matter whether the observer is stationary or moving. What is important is determining the moment of measurement with a third observer, who watches the events closely and takes a picture of every event he sees at the moment of its occurrence, so the picture is the witness that expresses the actual locations of events. The picture shows the time of the event and explains the events in detail.

To achieve this, we use the following thought experiment:

At a specific moment, solar explosions occurred, and at the same time, a rocket was launched from Earth toward the sun at a speed of one-quarter the speed of light. When Venus is located between Earth and the Sun, the rocket heading from Earth will reach Venus in six minutes, the same time it takes for light from the Sun to reach Venus since the moment of the solar explosions. In this description, the time of the moving frame is equal to that of the light ray coming from the explosion. The observer in the rocket is the first to see the solar explosions six minutes after they occurred. This is the second image, as the first image depicts the solar explosions occurring six minutes earlier, coinciding with the rocket's launch from Earth.

As for the observer on Earth, the image of the solar explosions will reach them eight minutes after they occur, which is the third image. Here, we find that the difference in the time of viewing by the moving observer precedes the observation of the event by the stationary observer by two minutes. During these two minutes, the rocket will have advanced toward the Sun, a distance equal to the rocket's speed multiplied by the time it took for light to traverse the distance between Venus and Earth.

Here, we observe that we are dealing with the ray emitted from the event so that each observer will see the event according to their location and in their own time. The ray from the event intersects with the moving frame, each covering a distance determined by the frame's speed. We can clarify it as follows:

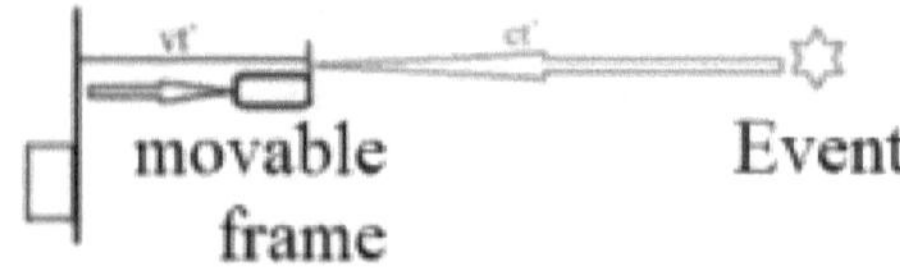

We can derive the transformations based on the data of the above thought experiment, as shown below:

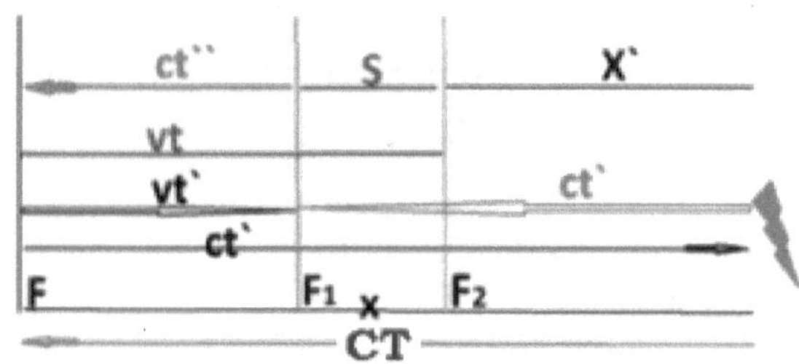

Where:

F: Position of the stationary frame.

F`₁: The first position for the arrival of the moving frame.

F`₂: The second position for the arrival of the moving frame.

R: The event (explosion).

X`: The distance between the moving frame and the event.

X: The distance between the stationary frame and the event.

S: The distance covered by the moving frame during the time it takes for the image of the explosion to move from the moving frame to the stationary observer.

c: The speed of light.

v: The relative velocity between the two frames.

t: Time of the stationary frame.

t`: Time of the moving frame.

t``: Time of light to traverse the distance from the moving frame to the stationary frame.

T: Time of light to traverse the distance from Event to the stationary frame

According to the order of events, the first event was an explosion at the event's site. At that exact moment, the observer in the moving frame moved from the match's location between the two observers toward the site of the explosion.

The second event is that the moving observer arrived at the first location, where he traveled a distance of vt`.

At the moment of his arrival, he saw the image coming from the explosion, which was now at a distance of ct`.

Here, we find that the time of the beam is equal to the time of the moving frame, and we can compensate for the distance travelled by the

beam. This is the difference between these new transformations and the Lorentz transformations.

The derivation of the transformation equations becomes:

The distance that the moving frame moves, during the period of time when the light beam travels, from the moving frame, until it reaches the resident observer, in order to achieve the observation, is:

$$ct'' = vt'$$

$$t'' = vt'/c$$

$$S = v.vt/c$$

$$S = v^2t/c$$

When the resident observer sees the explosion, at that moment, he will see the moving frame in the first location, while he has reached the second location. The actual total distance at the moment of viewing is the distance travelled by the moving frame, which is the distance times time, added to the distance to which he moved, until the resident observer saw the event, as in the following equation:

$$vt = vt' + S$$

$$vt = vt' + v^2t/c$$

$$vt = vt'(1+v/c)$$

$$vt' = vt*\frac{1}{(1+v/c)}$$

$$t' = t*\frac{1}{(1+v/c)}$$

In the moment of measurement, the correction factor for distances to become realistic becomes:

$$\Sigma_{=} \frac{1}{1+v/c}$$

The third event is when the passive observer witnesses the explosion. At that moment, he will see the moving frame in the first location while it has arrived at the second location, so the actual total distance at the moment of viewing will be the visible distance plus the unseen distance, as in the following equation:

$Vt = vt` + S$

$$Vt = vt . \frac{1}{1+v/c} + v^2 t/c . \frac{1}{1+v/c}$$

Then, the realistic transformations become in the following mathematical form: after adding the unseen distance and multiplying the distance travelled in the equation by the correction factor:

$$x` = x - (vt + v^2 t/c)\, \frac{1}{1+v/c}$$

$$x` = x - \left(\frac{vt}{1+v/c} + \frac{v^2 t/c}{1+v/c} \right)$$

The total distance at the moment a resident observer sees the moving frame is the following equation:

$$x = x` + vt` + v^2 t`/c$$

The last equation does not need a correction coefficient, because the equation is used by the static observer.

To test the three transformations, we conduct a practical experiment in which the moving frame travels at half the speed of light toward the event in one second, and we analyse the results by applying the experiment to the three transformations, where we consider that the distance between the stationary frame and the event is one metre, and the light travels it in one second.

1 - Galileo's transformations

$$x` = x - vt$$

An observer watching the experiment from his observatory on Mars can use Galileo's transformations because the light rays from the stationary frame, the moving frame, and the event are all in one frame and the same distance away from it. Then, the results will be as follows:

$$x` = 1 - 0{,}5*1 = 0{,}5 \text{ mc}$$

2 - Lorentz transformations

$$x` = \dfrac{\dfrac{x-vt}{\sqrt{1-v^2/c^2}}}{}$$

$$x` = \dfrac{1-(0{,}5*1)}{\sqrt{1-0{,}5^2/1^2}} = 0{,}57735 \text{ mc}$$

Determining a clear meaning expressed by the number resulting from the equation is impossible. It is not the actual distance, nor is it the perceived distance, nor is it the location of the moving frame.

3 – Sadik Transformation

$$x` = x - vt.\dfrac{1}{1+v/c} - \dfrac{v^2 t}{c} \cdot \dfrac{1}{1+v/c}$$

$$x` = 1 - 0{,}5.\dfrac{1}{1+0{,}5/1} - \dfrac{0{,}25.1}{1}.\dfrac{1}{1+0{,}5/1} = 0{,}5 \text{ mc}$$

This result is the actual distance for all observers.

When we talk about motion, especially when speeds are close to the speed of light, we are talking about the actual locations of events as perceived by observers, not just how we see them. For example, when we look at the moon, at the moment of observation, 1.27 seconds have passed, which is the time it takes for the moon's light to reach us. During this time, the moon has moved at a distance of about 1.4 metres. As a result, the actual location differs from the observed location because what we see in reality is an image coming from the past, no matter how far the object we are observing is. Even if it is only one metre away, we still see an image that occurred one part 299,792,458 parts a second ago.

This description of the motion of moving objects and determining their actual locations aligns with the principle of relativity, which states that the physical laws remain unchanged when used by all observers, regardless of their velocities.

As we can describe, the length of the moving body in a mathematical equation, according to the new conversions, is as follows:

$$L = \frac{vt}{1+v/c} + \frac{vT}{1+v/c}$$

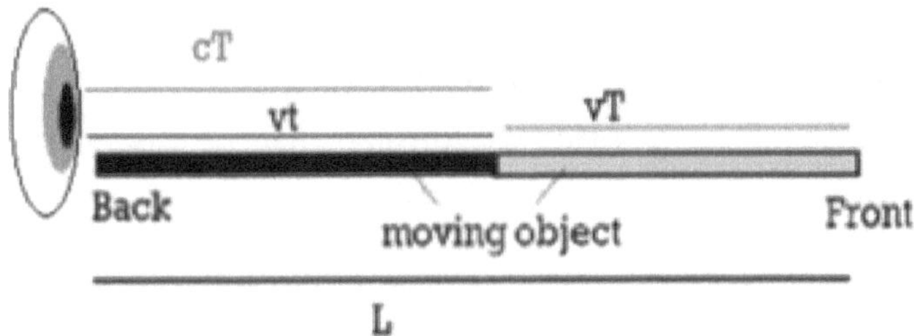

The actual distance between the front and the back of the body, is two distances, at the moment of viewing. The Introduction shall have been made in two stages, in two different and consecutive times. While the observer, he will only watch the first phase of the movement. Movement can be described as the following equations:

Where:

L: Mobile Body Length

V: Moving body speed

T: Moving body time from the beginning of motion to the moment of measurement

T: the time it takes for light to reach the observer

C: Speed of Light

Vt: body length as seen by the observer.

vT: the Distance You Travel Provided During Beam Arrival Time

cT: the distance the light travels from the introduction to the controller.

The time taken by the beam of light coming from the front to the observer is equal to the distance travelled, divided by the speed of light:

$T=vt/c$

In order to get rid of the difference in the distance between the two ends of the moving object, we consider that the observer is located in the location of the nearby party, so that the equation becomes as follows:

$$L = vt. \frac{1}{1+v/c} + \frac{v^2 t}{c} \cdot \frac{1}{1+v/c}$$

$L = 50 + 50 = 100$ cm

The equation expresses the length of a moving object that distances from the observer, the same length as a resident, although the observer considers the object to be only 50 centimetres long. This confirms that viewing does not reflect actual reality.

The equation that describes viewing the length of an object, approaching the observer, is different from the equation of viewing the length of the object that distances from the observer, and can be formulated as follows:

$L=cT-vt$

Below is a table of distances for various speeds.

Time	Light	Distance	Vilocity	Trans.D	Factor	vt`	s	x`	Distance
t	C	d = x	vC	vt	1/(1+v/c)	vt.ħ	(v²t/c).ħ	x-vt`-s	x=vt`+ct`
t	C	d	V	D	ħ	x=1	x=1	x=1	x
1	1	1	0,000001	0,00000	1,00000	0,000001	0,000000	0,999999	1,000001
1	1	1	0,000928	0,00093	0,99907	0,000927	0,000001	0,999072	1,000928
1	1	1	0,100000	0,10000	0,90909	0,090909	0,009091	0,900000	1,100000
1	1	1	0,200000	0,20000	0,83333	0,166667	0,033333	0,800000	1,200000
1	1	1	0,250000	0,25000	0,80000	0,200000	0,050000	0,750000	1,250000
1	1	1	0,300000	0,30000	0,76923	0,230769	0,069231	0,700000	1,300000
1	1	1	0,366025	0,36603	0,73205	0,267949	0,098076	0,633975	1,366025
1	1	1	0,450000	0,45000	0,68966	0,310345	0,139655	0,550000	1,450000
1	1	1	0,500000	0,50000	0,66667	0,333333	0,166667	0,500000	1,500000
1	1	1	0,600000	0,60000	0,62500	0,375000	0,225000	0,400000	1,600000
1	1	1	0,618034	0,61803	0,61803	0,381966	0,236068	0,381966	1,618034
1	1	1	0,650000	0,65000	0,60606	0,393939	0,256061	0,350000	1,650000
1	1	1	0,707107	0,70711	0,58579	0,414214	0,292893	0,292893	1,707107
1	1	1	0,750000	0,75000	0,57143	0,428571	0,321429	0,250000	1,750000
1	1	1	0,800000	0,80000	0,55556	0,444444	0,355556	0,200000	1,800000
1	1	1	0,866025	0,86603	0,53590	0,464102	0,401924	0,133975	1,866025
1	1	1	0,900000	0,90000	0,52632	0,473684	0,426316	0,100000	1,900000
1	1	1	0,990000	0,99000	0,50251	0,497487	0,492513	0,010000	1,990000
1	1	1	0,999900	0,99990	0,50003	0,499975	0,499925	0,000100	1,999900
1	1	1	0,999990	0,99999	0,50000	0,499997	0,499993	0,000010	1,999990
1	1	1	0,999999	1,00000	0,50000	0,500000	0,499999	0,000001	1,999999
1	1	1	1,000000	1,00000	0,50000	0,500000	0,500000	0,000000	2,000000

Thanks to new transfers, every observer is in a moving frame. However, he feels that in the context of an absolute inhabitant, which is his reference for monitoring, if the beam is exported or coming to him, the result is unchanged for him. However, we cannot compensate for one beam instead of the other, because the distance travelled by each beam varies, depending on the time taken by each of them, as the beam coming from the moving frame has emanated from the moment of conformity with the static frame. The beam from the event goes the remainder of the distance, which we symbolize with the symbol " X`

Relativity principle

The principle of relativity confirms that all the physical laws used by an observer at rest remain the same without any change when used by a moving observer within a frame of self-consistency. Based on this principle, we can transform the readings of the two observers because they are both using the same law. The connection between the two observers' readings is called "transformations".

Since the relationship is relative between frames, and both frames are in motion, we, as a third observer, cannot determine the velocity of either frame relative to the actual and absolute rest frame in the universe. We call this the "etheric velocity," and the reason is that we are also within a moving frame.

Based on the results of the Michelson and Morley experiment, which proved that light travels at the same speed in all directions for an observer on the ground, even though the ground is moving, this confirms that the speed of light is variable, for each observer, and does not belong to the same frame in which the light moves. Based on the stability of the physical law for all observers, nothing prevents any observer from using all the physical laws as they are without any change in their mathematical form, including Maxwell's equations in electromagnetism.

This confirms that the laws do not change with speed. What applies to the Earth, which revolves around the sun at a speed of 30 kilometres per second, applies to any frame, regardless of its relative or ethereal speed.

This result is what achieves the stability of all the laws of physics in different frames of reference when any observer uses them in his own frame, as there is no difference between the observers, whom we assumed in a static frame, or the observer in a moving frame because the laws apply to all observers equally.

Thus, we find that every frame in the universe is moving within another moving frame. Every material that falls under the influence of the frame's movement acquires its speed and direction, inside and outside the frame. As for sound and light, they are waves that move with the frame and are affected by their movement and speed only inside the frame, and when they move outside the frame, they are not related to its speed and direction. It has different values for all observers according to speed and direction.

This is confirmed by the experiment by the scientist Fizeau, which states *that a moving liquid partially drags light with it when light transitions from air to liquid.*

According to the concept of "Einstein" in relativity, when we talk about a moving object, and we want to know the impact of speed on its physical properties, we must apply all the special relativity equations to it in one push, and no equation may be applied, and the rest dispensed with. So, we find that physical properties will remain constant and do not change with speed change. We can easily confirm this by conducting a one-metre train car mental experiment that moves at half the speed of light. How long does it get and how long will it slow down?

$$L = L_0 / \dfrac{1}{\sqrt{1 - v^2/c^2}}$$

$$L = 1 * \sqrt{0,75}$$

$$L = 0,866 \ m$$

Where:

L - Train carriage length during movement

L_0 - Length of train carriage in case of stillness

V - Train Speed

C - Speed of light inside the train carriage

This means that the carriage contracted from 1 metre to 0.866 metres.

To measure the time that has slowed down in the carriage at full length without contraction, we find it in the following formula:

$$T = t^* \frac{1}{\sqrt{1 - v^2/c^2}}$$

T = 1/0,866

T = 1,1547344 Sec

With this amount of time, the light travels one metre inside the carriage during the movement.

Where:

T - Time in the carriage

t. Time within the residential observer

To measure the amount of time required to cover a distance of 0.866 metres, which is the length of the deflated cart, during movement. We multiply the distance by the time required to travel one meter, as follows:

1,1547344* 0,866 = 1Sec

Here, when the relative speed of a moving object causes a contraction in the body's length, time in the moving frame becomes identical to the observer's time in the static frame. If we write the equations vice versa, we will find that in the case of movement, the length remains constant and congruent to the length in the case of stillness. This means that the theory of relativism is true, the requirement that all equations apply simultaneously, which is what nobody paid attention to.

Thus, the Michelson-Morley experiment paved the way for the derivation of the comprehensive, unified theory of relativity, which is the theory that explains why the physical law is constant in all frames of reference without the need to convert between observers' readings, and there is no need for all theories of relativity.

Postulates (axioms) of relativity theory

1. The laws of relativity are limited only to objects subject to the law of inertia.
2. It is impossible for any two observers in two different frames to agree on the speed and direction of any third frame.
3. Every frame in the universe is moving. However, every observer has the right to claim that he is in a static frame, theoretically and mathematically, which qualifies him to use all physical laws, including the laws of electromagnetic theory.
4. All physical laws a stationary observer uses remain constant without any change when used by a moving observer, regardless of his speed.
5. The speed of light is constant and is related to the speed of the frame when the light moves inside the frame. At the same time, the speed of light varies relative to any observer in any other frame, each according to its movement's direction and relative speed with the frame in which the light moves. The relationship between the frame and the light ends when the light leaves the frame into a second frame.

Theory of Third Relativity

Applying the five principles, we can dispense with all the transformation equations and replace them with a single equation that applies to all observers equally and without exception. Stillness can either be absolute or not. We have no right to consider a body stationary while it is moving, while any observer is the only one who has the right to claim that they are in a stationary frame.

The theory aligns perfectly with the principle of relativity, affirming the constancy of the physical law that now has a clear scientific justification and explanation. There is no longer a need for new transformations, not just Lorentz transformations but even Galilean transformations are no longer necessary.

We can describe this scenario as follows:

someone throws a ball from the front of the train towards the back of the train at the same speed as the train. In this case, the ball becomes stationary relative to the train tracks, as well as for the observer standing at the train station. The train carriage is the one moving relative to the ball, just as it moves relative to the tracks. All we need to do for this is to change the labels and combine the two Galilean transformation equations into one equation, which is:

$$x = x' + vt$$

So, all the physical laws used by one observer are the same as those used by the second observer, without any changes, or transformations. The difference lies only in the direction of motion, which is unrelated to the mathematical outcome of the equation.

Where:

Vt: Distance between the two observers as one of them moves away from the other.

X`: The changing distance between the observer and the event.

X: The constant distance between the observer and the event.

The equation applies to any object, even if it is moving at the speed of light. It also applies to light moving within the frame. In this case, light is always the one in motion, and the observer is at rest, and it cannot be the other way around. The equation represents actual rather than apparent positions.

With this theory, the constant speed of light in all directions relative to the Earth has been explained. As long as the observer, the measuring device, and the light are all in the same reference frame, which is the Earth, the speed of light remains constant. However, the perception of events may differ when the observer is in a different reference frame.

Using this theory, the constant speed of light in all directions relative to the Earth has been explained. As long as the observer, the measuring device, and the light are all in the same reference frame, which is the Earth, the speed of light remains constant. However, the perception of events may differ when the observer is in a different reference frame. Although the motion of the ball itself will not change, it is the frame of observation that alters. Therefore, we find that observation does not represent the actual and true reality of the event, nor does it reveal the direction of motion or the actual speed. Each observer perceives the event differently, depending on their own velocity and direction.

Don't miss out!

Visit the website below and you can sign up to receive emails whenever Hasan Sadik publishes a new book. There's no charge and no obligation.

https://books2read.com/r/B-A-SSPU-YUUPC

BOOKS2READ

Connecting independent readers to independent writers.

About the Author

The author was born and raised in Baghdad, Iraq. After completing his secondary education, he joined the Physics Department at Damascus University. After graduating, he returned to Iraq and worked in the industrial and commercial sectors. In 1995, he moved to Libya, where he worked for oil companies. In 2000, he travelled to Germany and worked in various business ventures, eventually obtaining German citizenship. Due to his passion for knowledge and his interest in physics, after retiring from work in 2017, he returned to reading and studying, conducting scientific research, and publishing it in specialised scientific journals. In 2021, the author published his first book titled "The Shock of Physics in the Battle of Geniuses", which is the culmination of his research efforts.

Read more at https://www.facebook.com/groups/616359475891883/.

www.ingramcontent.com/pod-product-compliance
Lightning Source LLC
Chambersburg PA
CBHW031342160726
47993CB00002B/793